Medical Treatment of Intoxications and Decontamination of Chemical Agents in the Area of Terrorist Attack

NATO Security through Science Series

This Series presents the results of scientific meetings supported under the NATO Programme for Security through Science (STS).

Meetings supported by the NATO STS Programme are in security-related priority areas of Defence Against Terrorism or Countering Other Threats to Security. The types of meeting supported are generally "Advanced Study Institutes" and "Advanced Research Workshops". The NATO STS Series collects together the results of these meetings. The meetings are co-organized by scientists from NATO countries and scientists from NATO's "Partner" or "Mediterranean Dialogue" countries. The observations and recommendations made at the meetings, as well as the contents of the volumes in the Series, reflect those of participants and contributors only; they should not necessarily be regarded as reflecting NATO views or policy.

Advanced Study Institutes (ASI) are high-level tutorial courses to convey the latest developments in a subject to an advanced-level audience

Advanced Research Workshops (ARW) are expert meetings where an intense but informal exchange of views at the frontiers of a subject aims at identifying directions for future action

Following a transformation of the programme in 2004 the Series has been re-named and re-organised. Recent volumes on topics not related to security, which result from meetings supported under the programme earlier, may be found in the NATO Science Series.

The Series is published by IOS Press, Amsterdam, and Springer, Dordrecht, in conjunction with the NATO Public Diplomacy Division.

Sub-Series

A. Chemistry and Biology	Springer
B. Physics and Biophysics	Springer
C. Environmental Security	Springer
D. Information and Communication Security	IOS Press
E. Human and Societal Dynamics	IOS Press

http://www.nato.int/science
http://www.springer.com
http://www.iospress.nl

Series A: Chemistry and Biology – Vol. 1

Medical Treatment of Intoxications and Decontamination of Chemical Agents in the Area of Terrorist Attack

edited by

Christophor Dishovsky
Military Medical Academy,
Sofia, Bulgaria

Alexander Pivovarov
Ukrainian State University of Chemical Engineering,
Dnepropetrovsk, Ukraine

and

Hendrik Benschop
TNO Prins Maurits Laboratory,
Rijswijk, The Netherlands

Springer

Published in cooperation with NATO Public Diplomacy Division

Proceedings of the NATO Advanced Research Workshop on
Medical Treatment of Intoxications and Decontamination of Chemical Agent in the Area of Terrorist Attack
Dnepropetrovsk, Ukraine
25-28 January 2005

A C.I.P. Catalogue record for this book is available from the Library of Congress.

ISBN-10 1-4020-4169-1 (e-book)
ISBN-13 978-1-4020-4169-3 (e-book)

Published by Springer,
P.O. Box 17, 3300 AA Dordrecht, The Netherlands.

www.springer.com

Printed on acid-free paper

Printed in the Netherlands

CONTENTS

PREFACE

This book includes reports which were presented at a NATO Advanced Research Workshop (ARW) entitled "Medical Treatment of Intoxications and Decontamination of Chemical Agent in the Area of Terrorist Attack". The workshop was held on 25-28 January 2005 in Dnepropetrovsk, Ukraine.

Currently, scientists and experts from many countries are working on development and implementation of a readiness for anti terrorist actions. In addition to chemical weapons, terrorists can use various toxic chemicals from chemical industry, agriculture or products released from terrorist acts on industrial facilities. The arsenal of chemical agents that can be used as terrorist agents is practically unlimited. The focus of this workshop was assessment of scientific concepts and practical means for management of chemical agent casualties in the area of terrorist attack with emphasis on improving the medical treatment and decontamination. These problems were analyzed from an interdisciplinary perspective.

The primary objective of the ARW was to provide countries in which chemical-defense systems are currently under development with the experience and expertise of those that are more advanced and/or have already been exposed and responded to a chemical-terrorist attack. .

The main areas covered were:

1. new approaches in pre-treatment and prophylaxis of nerve agent intoxication;
2. diagnosis of exposure to chemical agents ;
3. therapy of chemical agent intoxication;
4. development of personal decontamination;
5. decontamination of intoxication with chemical agents.

ACKNOWLEDGEMENTS

The meeting was sponsored by the NATO Scientific and Environmental Affairs Division. We take this opportunity to express our thanks to the sponsor. We also are grateful to Dr. Sergey Ryzhenko for according a hearty welcome at Dnepropetrovsk Region Sanitary-Epidemiological Station, Dr. Grigory Shmatkov for a very interesting and useful excursion to the industrial city of Dneprodzerzhinsk , Mr. Valery Berestjanoy for organizational contacts with the Dnepropetrovsk Regional Administration. Finally, the hard and dedicated work of Dr. Viktor Kovalenko, Dr. Anna Tischenko, Mrs. Tetjana Bolva, Mr. Boris Dubina, Ukrainian State University of Chemical Engineering, Dnepropetrovsk, is highly appreciated. Indeed, without their organizational and technical support this meeting would not have gone as smoothly as it did.

Prof. Christofor Dishovsky, Ph.D., D.Sc. – Co-Director
Prof. Alexander Pivovarov, D.Sc. – Co-Director

KEY SPEAKERS

Dishovsky Christophor, Ph.D., D.Sc., Professor.
Military Medical Academy, Department of Military Toxicology.
3,St.G.Sofiisky Str. Sofia 1606, Bulgaria
Telephone: (+359 2) 9226448,
Fax: (+359 2) 952 6536
E-mail: christophord@yahoo.com

Pivovarov Alexander, D.Sc., Professor.
Ukrainian State University of Chemical Engineering, Department of Food Equipment and Technology. 8 Gagarin Ave.,
Dnepropetrovsk 49005, Ukraine
Tel. (+38 0562) 38-56-58
Fax: (+38 0562) 36-68-37
E-mail: apivo@ua.fm

Benschop Hendrik, Ph.D., Professor.
Department of Chemical Toxicology, TNO Prins Maurits Laboratory,
2280AA Rijswijk, Postbox 45.The Netherlands.
Tel. 00 31 15 2843629
Fax: 00 31 15 2843963
E-mail: benschop@pml.tno.nl

Philippens Ingrid H.C.H.M., D.Sc.
Department of Chemical Toxicology, TNO Prins Maurits Laboratory,
2280AA Rijswijk, Postbox 45.The Netherlands.
Tel. 00 31 15 2843048
Fax: 00 31 15 2843963
E-mail: philippens@pml.tno.nl

Jiri Matoushek, Professor,
Masaryk University, Faculty of Sci., EU Res. Ctre of Excellence for Environmental Chemistry and Ecotoxicology, Kamenice 126/3, CZ-625 00 Brno, Czech Republic
Tel. 420 549 492 860
Fax: 420 549 492 840
E-mail: matousek@recetox.muni.cz

Jyruli Manana, D.Sc.
Georgian Environmental and Biological Monitoring Association.
60 Agmashenbeli Av., Tbilisi, 0102 Georgia
Tel. +99532 98 67 94
Fax: +99532 95 67 92
E-mail: mjuruli@myoffice.ge

Brown Clarence Lee, D.Sc.
Inspection Team Leader / Senior Medical Officer Inspectorate, Organization for the Prohibition of Chemical Weapons (OPCW), Johan de Wittlaan 32 2517 JR Den Haag. The Netherlands
Tel. +31 (0)70-416.31.74
Fax: +31(0)70-306.35.35
E mail: Clarence.Brown@opcw.org

Rowell Mike, MB,BS,MPH
Senior Medical Officer Health and Safety Branch, Organization for the Prohibition of Chemical Weapons (OPCW), Johan de Wittlaan 32 2517 JR Den Hague. The Netherlands
Tel. +31 (0)70-416.35.10
Fax: +31 (0)70-306.35.35
E-mail: mrowell@opcw.org

Ryzhenko Sergey, D.Sc. (Chief of DRSES)
Dnepropetrovsk Region Sanitary-Epidemiological Station.
39a. Philisofskaja St., Dnepropetrovsk 49000, Ukraine
Tel. +38 056 3710155
Fax: +38 056 3710155
E-mail: oblses@a-teleport.com
http://www.sesobl.dp.ua

Khaydarov Renat, D.Sc.
Institute of Nuclear Physics, 702132, Ulugbek, Tashkent, Uzbekistan, (998712)606144
Tel. 998 712 606144
Fax: 998 712 606144
E-mail: physicist@sarkor.uz

Litvin Yuri, Professor., Col.
Dnepropetrovsk State Medical Academy, Chief of Department of Extremely and Military Medicine. 9. Dzerzhinskogo St. 49044 Ukraine
Tel. +38 0562 938916
Fax: +38 056 7702258
E-mail: litvin@dsma.dp.ua

Popov Tzvetan, D.Sc.
State Agency for Civil Protection, 30 Gabrovsky St., Sofia 1172, Bulgaria.
Tel. 00359 2 960 10 324
Fax: 00359 2 960 10 324
E-mail: tsvtnppv@abv.bg

Zadorsky Willjam, Professor.
Ukrainian State University of Chemical Engineering,
Department of Chemical Equipment and Ecology. 8 Gagarin Ave.,
Dnepropetrovsk 49005, Ukraine
Tel. +38 (056) 744 0210
Fax: +38 (056) 744 0210
E-mail: ecofond@ecofond.dp.ua
http://zadorsky.com

Kokshareva Natalia, Professor.
Medved's Institute of Ecohygiene and Toxicology,
6. Heroev Oborony St. 01000 Kiev, Ukraine
Tel. +38 (044)258 32 01
Fax: +38 (044)251 96 43
E-mail: knv@medved.kiev.ua

Tkach Vladimir, Professor.
Medved's Institute of Ecohygiene and Toxicology,
6. Heroev Oborony St. 01000 Kiev, Ukraine
Tel. +38 (044)258 32 01
Fax: +38 (044)251 96 43
E-mail: knv@medved.kiev.ua

Shmatkov Grigoriy, Professor.
Dnepropetrovsk Regional Administration, 1. Kirov Av.,
Dnepropetrovsk 49004, Ukraine
Tel. +38 (0562) 428841
Fax: +38 (0562) 424028
E-mail: shmatkov@gue.dp.ua

Florin Paul, MD, Ph.D., Col.
Str. Institutul Medico Militar NR 3-5, Bucharest, Romania
Tel. +40-21-212-6004
Fax: +40-21-212-6004
E-mail: paulf@pcnet.ro

Pozdnjakova Ludmila, MD, Ph.D.
Ukrainian I.I.Mechnicov Anti Plague Research Institute,
Odessa 65000, Ukraine.
Tel. +38 (048) 741 1586
E-mail: PL258396@ua.fm

OTHER PARTICIPANTS AND OBSERVERS

Kovalenko Victor, D.Sc. (Pro-rector)
Ukrainian State University of Chemical Engineering,
8 Gagarin Ave., Dnepropetrovsk 49005, Ukraine
Tel. +38 (056) 744 0210
Fax: +38 (056) 744 0210
E-mail: ughtu@dicht.dp.ua

Hmel Stanislav, Col.
Dnepropetrovsk State Medical Academy, Department of
Extremely and Military Medicine. 9. Dzerzhinskogo St.
49044 Ukraine
Tel. +38 0562 938916
E-mail: litvin@dsma.dp.ua

Gninenko Alexander, Lieut. Col.
Dnepropetrovsk State Medical Academy, Department of
Extremely and Military Medicine. 9. Dzerzhinskogo St.
49044 Ukraine
Tel. +38 0562 938916
E-mail: litvin@dsma.dp.ua

Kremenchutzky Gennady, Professor.
Dnepropetrovsk State Medical Academy, Chief of
Department of Microbiology. 9. Dzerzhinskogo St.
49044 Ukraine
Tel. +38 0562 471355
E-mail: kremen@dsma.dp.ua

Bandura Alexey, Captain.
Science-Research Centre of Expert Crime Detection.
Chief of Research Explosive Substances Department of
Police in Dnepropetrovsk Region. 105a, Heroev Stalingrada St.,
Dnepropetrovsk 49033, Ukraine
Tel. +38 0562 415798

Isaev Viktor, MD. Ph.D. (Deputy of Chief).
Dnepropetrovsk Region Sanitary-Epidemiological Station.
39a. Philisofskaja St., Dnepropetrovsk 49000, Ukraine
Tel. +38 056 3710155
Fax: +38 056 3710155
E-mail: oblses@a-teleport.com
http://www.sesobl.dp.ua

Isaev Andrey, MD. Ph.D.
Dnepropetrovsk Region Sanitary-Epidemiological Station.
Chief of Sanitary-Hygienic Department.
39a. Philisofskaja St., Dnepropetrovsk 49000, Ukraine
Tel. +38 056 3710155
Fax: +38 056 3710155
E-mail: oblses@a-teleport.com
http://www.sesobl.dp.ua

Berestjanoy Valery. (Deputy of Chief)
Main Directorate of External Economic Links of
Dnepropetrovsk Regional Administration,
1. Kirov Av., Dnepropetrovsk 49004, Ukraine
Tel. +38 (0562) 428209
Fax: +38 (056) 74412426
E-mail: isachenko@adm.dp.ua

Kolyada Alexander, MD. Ph.D.
Medical Academy of Postdiploming Education,
20 Vintera Av. Zaporizhia 69096, Ukraine
Tel. +38 (0612) 570533
Fax: +38 (0612) 570533
E-mail: iskander@mail.zp.ua

Andruschenko Nikolai, Col. (Deputy of Chief)
Department of Extraordinary Situation and Protection
of Population from Consequences of Chernobyl Accident
in Dnepropetrovsk Region. 4 Korolenko St. Dnepropetrovsk
49000 Ukraine.
Tel. +38 (0562) 7701891

Tischenko Anna, D.Sc.
Ukrainian State Chemical-Technology University,
Department of Food Equipment and Technology.
8 Gagarin Ave., Dnepropetrovsk 49005, Ukraine
Tel. (+38 0562) 470555
Fax: (+38 0562) 356422
E-mail: antisch@ua.fm

Kharytonov Mykola, D.Sc.
Ecotoxicology Laboratory, Dnepropetrovsk State
Agrarian University, 25 Voroshilov St., Dnepropetrovsk,
49600, Ukraine
Tel. (+38 056) 7448132
E-mail: envteam@ukr.net

Part 1

CURRENT PROBLEMS OF CHEMICAL TERRORISM

Chapter 1

PROBLEMS OF CHEMICAL TERRORISM AND WAYS OF ITS OVERCOMING

Christophor Dishovsky

Department of Military Toxicology, Military Medical Academy, Sofia , Bulgaria

Abstract: The main problem connected with chemical terrorism is that, beside chemical weapons, terrorists can use different toxic chemicals from the chemical industry, from agriculture or products of industrial facilities released after the terrorist act. An attack to a chemical plant can instantly liberate a number of different chemicals. Studies should be made on incidents occurring in the facility, during transportation, storage or other processes and these are important points in the preparation for protection against chemical terrorism. An important corner-stone of the anti-terrorist organization in any country is to set a Health & Disaster/Anti Terrorist Acts Management system. Country antiterrorist protection will be improved by the introduction of a universal strategy on basic therapeutic trends against chemical terrorism. Chemical terrorism can be responsible not only for the spread of large amounts of toxic chemical compounds, but also for chronic and delayed effects of these agents. Intoxication with small doses of toxic agents is also a possibility used by terrorists. The variety of characteristics of a chemical agent used by terrorists needs demands improvement in the detection, personal protection and decontamination procedures, including that of the medical personnel and equipment. Antidote treatment, with the exception of the medical units and organization of national stockpiling which are adequately supplied for, needs a new and extensive study for new antidotes and for improvement of the medical treatment on the area of the terrorist act.

C. Dishovsky et al. (eds.), Medical Treatment of Intoxications and Decontamination of Chemical Agents in the Area of Terrorist Attack, 3–11.

Keywords: antidotes; Anti Terrorist Management Centers; chemical; chemical industry; chemical weapons; decontamination; delayed effects; detection; terrorism

1. INTRODUCTION

The main problem connected with chemical terrorism is that, beside chemical weapons, terrorists can use various toxic chemicals from the chemical industry, from agriculture or products of industrial facilities released after the terrorist act. An attack to a chemical plant can instantly liberate a number of different chemicals. Studies should be made on incidents occurring in the facility, during transportation, storage or other processes and these are important points in the preparation for protection against chemical terrorism. An important corner-stone of the anti-terrorist organization in any country is to set a Health & Disaster/Anti Terrorist Acts Management system. Country antiterrorist protection will be improved by the introduction of a universal strategy on basic therapeutic trends against chemical terrorism. Chemical terrorism can be responsible not only for the spread of large amounts of toxic chemical compounds, but also for chronic and delayed effects of these agents. Intoxication with small doses of toxic agents is also a possibility used by terrorists. The variety of characteristics of a chemical agent used by terrorists needs demands improvement in the detection, personal protection and decontamination procedures, including that of the medical personnel and equipment. Antidote treatment, with the exception of the medical units and organization of national stockpiling which are adequately supplied for, needs a new and extensive study for new antidotes and for improvement of the medical treatment on the area of the terrorist act.

2. DISCUSSION

The main problem connected with chemical terrorism is that beside chemical weapons, terrorists can use different toxic chemicals from the chemical industry, the agriculture or products released from industrial facilities following a terrorist act. An attack on a chemical plant can immediately release a number of different kinds of chemicals [6]. Some differences exist between chemical weapons (CW) and the chemicals released after destruction of a chemical plant following a terrorist act [4]:

- industrial chemicals are less toxic than CW, but will be present in much higher quantities for a longer period of time;
- contamination with the hazardous industrial chemicals eventually covers a bigger area;

- CW represent a relatively small number of potential agents; on the contrary - toxic industrial chemicals – tens of thousands;
- for the known CWs, relatively simple detection and identification equipment and methods have been developed; the potential variety of industrial chemicals makes the detection process very difficult;
- decontamination some times may be a long-lasting and expensive procedure;
- the products of the decontamination can also damage the environment;
- military protective filters are optimized against CW and BW; some hazardous industrial chemicals are not very well filtered by military filters;

The study of the accidents in chemical facilities, during transportation, storage and others, the research of the chemical products, which are produced and stored, are important points in the preparation of the defence against chemical terrorism. An example of such an accident is the explosion at the Union Carbide pesticide manufacturing plant (Dec. 3, 1984), which scattered toxic methyl isocyanate (MIC) over the city of Bhopal, India. During the first few days up to 4000 people died of painful, harrowing deaths [10].

The precise knowledge of the chemicals, produced and stored in the factories, is a necessary condition for modeling of the accidents. However, it should be taken into account that in real conditions and especially when the accidents are accompanied with fire, new toxic compounds may be formed. Such is the case with the fire, which occurred on July 13, 1993 in the "Alen Mak" factory in the town of Plovdiv, Bulgaria. During the GC-MS analysis of the air, soil and various parts of the incident site, over 120 different chemical compounds were identified, some of which identical to the ones present on the premises before the fire, while others had obviously formed during the process of combustion. For a great number of these compounds, toxicological data was not found in the accessible information banks, such as IRIS of EPA and others [5].

United Nations APELL (Awareness and Preparedness for Emergencies at the Local Level) [15] chemical accident database (storing information on about 300 major chemical accidents between 1970 and 1998) shows that:

- 38% of the accidents occur with hydrocarbons (fuel gases, fuel liquids, oil, or refined petroleum products);
- 15% - explosive industrial chemicals;
- 8% - chlorine;
- 6% - ammonia;
- 6% - industrial acids and bases;
- 3% - pesticides and chemical intermediates;

- 21% - polychlorinated biphenyls (2 %), unspecified chemicals (5%) and others.
Hydrocarbon production, storage, transportation and distribution facilities are at the top of the list of potential targets.

Some of the more common types of chemicals that could be used as improvised weapons against a community include (data from *ATSDR* [16]:
- eye, skin and respiratory irritants (acids, ammonia, acrylates, aldehydes, and isocyanates);
- choking agents (chlorine, hydrogen sulfide, and phosgene);
-flammable chemical industry gases (acetone, alkenes, alkyl halides, amines);
- aromatic hydrocarbons that could be used as water contaminants (benzene, etc.);
- oxidizers for improvised explosives (oxygen, butadiene, and peroxides);
- aniline, nitrile, and cyanide compounds that could be used as chemical asphyxiants;
- compressed hydrocarbon fuel gases that could be used as incendiaries or simple asphyxiants (liquified natural gas, propane, isobutane);
- liquid hydrocarbon fuels that could be used as incendiaries or water contaminants (gasoline, jet fuel);
- industrial compounds that could be used as blister agents (dimethyl sulfate), and
- organophosphate pesticides that could be used as low-grade nerve agents.

According to the conclusions reached by some experts, terrorist acts apply a new strategy in the use of chemical weapons and other chemical agents, i.e.:

- the use of low doses of toxic chemicals;
- to aim at a delayed and unknown effect of intoxication.

Investigations in Bhopal, 20 years later, showed that thousands of Bhopal residents suffered from serious long-term side effects such as blindness, and liver and kidney failure. Estimates of total deaths reach up to 20000. Injuries probably exceeded 100000 [10]. The data from terrorist acts in Matsumoto city and Tokyo underground in Japan are important for the understanding that the consequences of a terrorist act bring long lasting troubles [12, 13]. Some investigations showed that allied troops had been exposed to sarin during the first Gulf War and that might have caused the development of the "Gulf war syndrome" in some of the exposed soldiers [7]. The main problems caused by OPC are neuro-toxic delayed effects [8]. Shulga [14] suggested also the presence of delayed neuro-endocrine toxicity after intoxication with such compounds.

Planning and preparation of anti terrorist measures should be focused on the following specific points [4]:

- risk assessment for the use of chemical agents as terrorist agents with particular attention to toxic industrial chemicals and toxins;

- update assessment of the effective toxic levels that should cover both the known chemical weapons in view of the modern technologies of their use and toxic compounds and chemicals of industrial origin;
- inventory and assessment of the available means for medical treatment of chemical intoxications; assessment of the required amounts and types of antidotes (in view of the broader range of potentially toxic agents) and their update with development and introduction of new compounds;
- modernization and optimization of individual protection with particular focus on respiratory protection and protective clothing;
- creation of new National Pharmaceutical Stockpile, which ensures the availability of medicines, antidotes, medical supplies and medical equipment necessary to counter the effects of biological pathogens and chemical agents;
- creation of effective system of information and supply to the site of the terrorist act;
- assessment of the available means for indication and control of chemical contamination and the effectiveness of decontamination. It should include a broader range of potentially toxic agents and the available state-of-the-art technologies;
- acquires particular significance for the readiness of all levels of civil institutions and the army to counteract chemical terrorism. It should incorporate and implement the latest achievements of computer simulation and virtual reality technologies;
- intensive education and training of first responders and physicians is needed for meeting the medical challenges imposed by chemical and other weapons of terrorism.

A country's antiterrorist protection system should also incorporate a general State Strategy against chemical traumatism and terrorism [11]:

- medical chapter;
- universal program for diagnosis and treatment;
- organizational and medical program;
- social and informative chapter;
- social information regarding chemical traumatism;
- knowledge on chemical traumatism and terrorism.

Investigations of Wetter et al. [17] showed that hospital emergency departments generally are not prepared to treat victims of chemical or biological terrorism. Some countries have introduced a Health & Disaster/Anti Terrorist Acts Management system, which includes:

- Health & Disaster Anti Terrorist Management Centers. They can be a structure of already existing facilities – for example - of the Civil Defense. Some countries created new independent structures. Such centers could act independently, their task being to manage the situation after disasters or terrorist acts.
- Facilities for indication of toxic chemicals;

- Stationary centers for indication of toxic chemicals that work online and are connected to the Health & Disaster Anti Terrorist Management Centers.
- Mobile facilities for indication of toxic chemicals for investigating a potential area of a terrorist act.
- Regional Medical Centers. They can be Primary or Secondary – depending on the size and type of organization. These can be district hospitals with developed emergency units.
- Poison information centers. Regional Medical Centers can play this role.
- Mobile Medical units with Clinic for First aid treatment; Laboratory & Diagnostic units; operating theater; hospital, specifications for treatment of patients affected by chemical agents (indication, decontamination).
- Satellite communications;
- Independent energy supply;
- Mobile decontamination facilities;
- Training Centers (Center) for Mass Casualties Events.

This system will have a dual function. It will enhance not only emergency preparedness, antiterrorist activity but also the public health system. Such a view is suggested by Marmagas et al.[10] about possible connection between preparedness against biological terrorism, public health infrastructure, chronic disease and environmental health tracking network.

With a view to improve the level of security of the population from incidents and disasters, a Scientific Coordination Council was created in Bulgaria as an adjunct to the Permanent State Commission for Protection of the Population from Disasters and Incidents [5]. Attached to the Council are several expert committees which organize and conduct research and analysis, give expertise and make risk assessments for nuclear security, radiation, chemical and biological defense, emergency, seismic danger, meteorological and hydrologic problems and physical steadiness, suggest measures for securing and protecting objects which may add to the dangerous conditions. One of the first tasks of the Expert Committee for Chemical Defense was to identify the dangerous chemical conditions on the territory of Bulgaria. At present over 3400 chemical substances are produced in Bulgaria, among which mineral fetilizers, plastics, drugs, and others. The basic criteria for the assessment concern the quantity of reserve of industrially dangerous chemical compounds and compounds which form during fires or through the reactions of these compounds. The classification is also concerned with the existence of storage facilities, product conductors, transportation communications, and others, where risky situations, including terrorist actions, may be expected. The criteria for the above mentioned analysis may be defined as follows [5]:

- research of the object as a potential danger;
- analysis of the possible terrorist act;

- analysis of the available system for defense, security and protection;
- development of a project for improvement of a new system for defense, security and protection;
- organizing of the defense, security and protection of some chosen objects, and consideration and use of the experience;
- coordination of the efforts of the various departments, ministries and others in solving the problems of defense, security and protection of the potentially dangerous objects.

The country's antiterrorist preparedness will be improved with the introduction of a universal doctrine (strategy) suggested from Monov [11], with the basic therapeutic trends against chemical terrorism which includes mainly:

- antidotes;
- detoxication;
- reanimation-substitution and correction;
- immunobiological activity;
- antiviral and antibacterial activity;
- organoprotective activity.

A great number of severely injured people, after using organophosphorus compounds (OPC) like sarin for example, will need skilled medical help, enough antidotes, preparedness for decontamination and medical equipment such as breathing apparatuses or supplied air-line respirators. Some ideas exist how to solve these problems. In the Military Medical Academy, Sofia, a special antidote, which can be used after fires and a buddy aid antidote – for severe cases of intoxication were created. It can be introduced also as the first and buddy aid antidote. This antidote was prepared in ampoule form bearing the name NEMICOL 5T [9]. It is a multi component preparation which was tested with animals. The investigations show that this antidote very quickly recovers the breathing, the arterial pressure and the EKG in the intoxicated animals. The anticonvulsant activity was very strong and was demonstrated also with EEG experiments. Neuromuscular transmission blocked after OPC intoxication recovered 1-3 minutes after introduction of the Nemicol 5T. All investigated parameters stayed very stable.

Cowan et al. [1, 2] suggest the Multi-Threat Medical Countermeasure (MTMC) hypothesis – investigations for developing a single countermeasure drug with efficacy against different pathologies caused by multiple classes of chemical agents. Serine protease inhibitors can prolong the survival of animals intoxicated with the nerve agent soman and can also protect against vesication caused by the blister agent sulfur mustard. Poly (ADP-ribose) polymerase (PARP) inhibitors can reduce both soman-induced neuronal degeneration and sulfur-mustard-induced epidermal necrosis. Accordingly, the drugs with anti-inflammatory action against either nerve or blister agent might also display multi-threat efficacy for the inflammatory pathogenesis of both classes of chemical warfare agent [1,2].

3. CONCLUSION

Nowadays, terrorists have changed their tactics. Beside chemical weapons, they may use different toxic chemicals produced by the industry, the agriculture or failures in the industrial facilities during transportation, storage, fires and the like. They may take advantage also of the application of low doses of toxic chemicals which can provoke delayed and unknown effects of the intoxication.

Thorough studies of the chemical products, investigations of the incidents in chemical factories, storehouses and transportation dealing with chemical products are important points in the protection plan against chemical terrorism.

A country's antiterrorist preparedness should include also a State strategy against chemical traumatism and terrorism. The elaboration of Organization of Health & Disaster/Anti Terrorist Acts Management system is an important part of this preparedness.

The scientific research in the field of toxicology gains new dimensions and priorities. The introduction of a universal strategy for the basic therapeutic trends against chemical terrorism is necessary.

REFERENCES

1. Cowan F. M., Broomfield C.A., Lenz D. E., Smith W. J., Putative role of proteolysis and inflammatory response in the toxicity of nerve and blister chemical warfare agents: implications for multi-threat medical countermeasures. J. Applied Toxicology, 2003, 23, 3, 177-86.
2. Cowan B. S., Broomfield C. a., Stojiljkovic M. P., Smith W. J., A Review of Multi-Threat Mrdical Countermeasures against Chemical Warfare and Terrorism, Military Medicine, 2004, 169, 11, 850-55.
3. Dishovsky C., "The problems of Chemical Terrorism." In *Technology for Combating WMD Terrorism*, Peter J. Stopa, Zvonko Orahovec, ed's, NATO Science Series, Kluwer Academic Publishers, 2004.
4. Dishovsky C., "The problems of Chemical and Biological Terrorism." In *Medical Aspects of Chemical and Biological Terrorism – Biological Terrorism and Tramautism,* Alexander Monov and Christophor Dishovsky, eds, Publishing House of the Union of Scientists in Bulgaria, 2004.
5. Dishovsky C., Belokonsky I., Panchev N., The problems of defence in chemical industry-results of special investigation. Proceedings of the CB Medical Treatment Symposium Industry I; 1998 October 25-31; Zagreb-Dubrovnik, Croatia, MOD of Croatia, 1999, 69-73.
6. Eifried G., Terrorism against chemical plants:hazards and risks. Proceedings of the CB Medical Treatment Symposium Industry I; 1998 October 25-31; Zagreb-Dubrovnik, Croatia, MOD of Croatia, 1999, 84-88.
7. Haley, R.W., and Kurt, T.R., Self-reported exposure to neurotoxic chemical combinations in the Gulf War, J. Am. Med. Assos., 1997, 277,231-37.

8. Haley, R., W., Billecke, S., S., and La Du, B., N., Association of low PON1 type Q (type A) arylesterase activity with neurological symptom complexes in Gulf War veterans, Toxicol. Appl. Pharmacol., 1999, 157, 227-33.
9. Kotev G., DSc Work , Military Medical Academy, Sofia, 1973. (in Bulgarian).
10. Marmagas S.W., Kind L. R.,Public Health's Response to a changed World: September 11, Biological Terrorism, and the Development of an Environmental Health Tracking Network. Amer. J. of Public Health, 2003, 93, 8, 1226-30.
11. Monov Alexander, "Biological Traumatism and Terrorism – unified Medical and Organizational Doctrine". ." In *Medical Aspects of Chemical and Biological Terrorism – Biological Terrorism and Tramautism,* Alexander Monov and Christophor Dishovsky, eds, Publishing House of the Union of Scientists in Bulgaria, 2004.
12. Morita, H., Yanagisawa, N., Nakajima, T., et al., Sarin poisoning in Matsumoto, Japan, Lancet, 1995, 346, 290-93.
13. Ohbu, S., Yamashina, A., Takasu, N., et al., Sarin poisoning on Tokyo subway. South. Med. J., 1997, 90, 587-93.
14. Shulga V., "Delayed neuro-endocrine toxicity indused by organophosphorus compounds- natural consequence of poisonous substances application for terrorist purpose", In *Medical Aspects of Chemical and Biological Terrorism –Chemical Terrorism and Tramautism,* Alexander Monov and Christophor Dishovsky, eds, Publishing House of the Union of Scientists in Bulgaria, 2005 (in press).
15. United Nations APELL (Awareness and Preparedness for Emergencies at the Local Level) http://www.uneptie.org/pc/apell/.
16. U.S. Department of Health and Human Services, Agency for Toxic Substances and Disease Registry (ATSDR) http://www.atsdr.cdc.gov/about.html.
17. Wetter D.C., Daniell W. E., Treser C. D., Hospital preparedness for victims of chemical or biological terrorism. Amer. J. of Public Health, 2001, 91, 5, 710-1

Chapter 2

TOXIC CHEMICALS AND RADIOACTIVE SUBSTANCES AS REASON OF OCCURRENCE OF ACUTE POISONING IN UKRAINE

Sergey Ryzhenko
Dnepropetrovsk Region Sanitary-Epidemiological Station.
39a. Philisofskaja St., Dnepropetrovsk 49000, Ukraine

The chemical terrorism is a new threat to mankind safety, vastly exceeding in scale the results of the most modern firearms and representing one of the cheapest forms of terrorism. The first information about "terrorism" dates from 431-404 years B.C., when ammonium smoke was used in the course of the Peloponness wars. In 1899 as a result of the International Peace Conference Agreement in Hague the use of artillery projectiles equipped with poisoning gas was prohibited. However it is in the XX century during World War I - in 1915 that the first large-scale use of poisonous material occurred on the battle site near the city of Ypres, Belgium. During World War I were applied more than 100 000 tons of toxic chemicals, as a result of which perished 90 000 soldiers and more than a million were affected.

In 1925 in Geneva a protocol was signed, which forbade the use of bacteriological and chemical weapons, however this was not sufficient for the countries to stop further production, use and accumulation of chemical weapons.

In 1972 several countries signed in Geneva a Convention on biological and toxin weapons and undertook the obligation to continue the negotiations for an agreement to forbid chemical weapons also. In 1992 the participants of the Geneva negotiations agreed on a text of the Convention on prohibiting the development, production, accumulation and use of chemical weapons and on their destruction. In 1993 the Convention on chemical weapons was open for signing.

During the ceremony for the Convention signing in January in Paris, 130 countries supported the agreement and international disarmament with their signatures. In February of the same year in Hague was founded a Starting-up commission with the task to prepare the coming into effect of the Convention. In spite of these efforts, in 1995 in Japan the “Aum Sinrike” sect made the terrorist act in Tokyo underground by using the chemical weapon sarin. Beside the 5000 affected, 12 persons perished.

C. Dishovsky et al. (eds.), Medical Treatment of Intoxications and Decontamination of Chemical Agents in the Area of Terrorist Attack, 13–19.

Only in 1997 the Convention on chemical weapons took effect for 87 member-countries – initiators of the Convention. According to the Convention to perform the requirements was created the Organization for Prohibition of Chemical Weapons (OPCW) with headquarters in Hague.

The Convention on Chemical Weapons is an international agreement, which forbids the use of chemical weapons and aims at liquidating the chemical weapons for ever and all over the world. The Convention provides the basis for supervision by the Organization for Prohibition of Chemical Weapons and for destruction of the available spare chemical weapons, as well as objects, which are used for production of chemical weapons, and for checking industrial objects for prevention not to allow the production of new chemical weapons. The Organization for Prohibition of Chemical Weapons also assists the international cooperation and exchange by research information so that people and governments use chemical investigations for peace purposes.

Modern highly toxic chemicals and biological agents could fall into terrorist hands through various channels:

- appropriation of military storehouses and arsenal, where chemical weapons are kept;
- appropriation of enterprises equipped with facilities for chemical protection;
- supplies from production, storage and trade companies;
- illegal production in underground laboratories.
- for the last 10-20 years there has been a qualitative growth in production of synthetic preparations for treatment of the various diseases, possessing a broad spectrum of pharmacological influence. Some psycho-pharmacological drugs can be used by terrorists for solving their own problems.

The sanitary and epidemiological service in Dnepropetrovsk area has functions, which have been entrusted by the legislation of Ukraine thus contributing to the preventive safeguarding of the people from poisoning by toxic chemicals applicable in industry and agriculture. For the industrial areas of Dnepropetrovsk the problem for protection from industrial wastes from enterprises is traditionally actual. At present over 8,2 billion tons of industrial wastes to miscellaneous degrees of toxicity are stored in this area. The wastes from the chemical production occupy more than 6,2 km^2. On the territory of the enterprise there continue to be accumulated wastes from the galvanic technology of 1st and 2nd degree of danger. The big worries remain the petroleum production waste, whose amount sharply increased in connection to the stormy development of the town public transport.

A requirement was introduced in this respect to undertake expert sanitary-hygienic measures for the treatment of wastes in accordance to article 24 of Law on Wastes of Ukraine, № 187-98-VR from 05.03.98 and articles 25 and 33 of Law of Ukraine "On the provision of sanitary and epidemic welfare for the population". Annually the sanitary and epidemiological service conducts over 500 expert operations. Till

01.01.2005 the sanitary-hygienic expert operations connected with wastes covered over 90% of the enterprises.

For the expert operations with wastes we used the official report form III for toxic wastes, issued by order of the Ministry of statistics of Ukraine on 23.07.93 under № 162. According to this form subject to accounting are all types of toxic wastes, including production unfit for recycled use and containing dangerous elements.

In order to elaborate a state system for registration of wastes, with Resolution 1360 from 31.08.98 the Council of Ministers of Ukraine approved the order for registration of sites where wastes are produced, recycled and used. With order № 41 from 17.02.99 the Ministry for environmental protection of Ukraine approved a form for register cards.

At present such register cards should be agreed on by the organs of state control of the district. The evaluation of the register cards is made on the basis of a sanitary-hygienic expertise of the wastes treatment. Special attention is paid to specified as being of high degree of danger wastes and their chemical composition in the register cards.

When making an expert evaluation, specialists at the sanitary and epidemiological service should take into account two types of waste storage sites: temporary and for a longer period of time. Temporary storage sites at the territory of the enterprise are given permission only by way of exception. Longer storage sites concern such installations as midding slime and slag collectors, mouldboards etc. Supervision of the sanitary and epidemiological service in this case is directed mainly to technical equipment requirements of the storage site and conducting selective laboratory monitoring of the atmospheric air, ground and water composition in the area of the wastes storage.

Annually the district uses hundreds of the tons of different pesticides (more than 100 different names) in agriculture. According to the Ukrainian sanitary regulation for transport and preservation of chemical preparations for plant protection, this activity can be performed only by transport vehicles and storehouses, which possess sanitary passports for ecological safety, issued by the State sanitary and epidemiological service and the Inspection for ecological safety. To work with pesticides are allowed only persons that have passed physical examination and have the corresponding education together with the special permit.

The rather thoughtless use of chemical compounds for plant protection in the second half of the XX century has brought about the accumulation in the storehouses of some facilities of about 1000 tons of preparations that are unfit for use, forbidden or unidentified. The sanitary and epidemiological service jointly with the Ecological inspection and the District station for plant protection are conducting an inventory of the storage sites of these toxic wastes. Inventory data with indications for the place of the storage, the names of the preparations and their quantity is kept in a computer database.

Funds have been assigned in the regional and local budgets for improving the storing of these toxic material, according to information from the sanitary-and-epidemiological service and the ecological inspection. So in 2004, about 150 tons of unfit and unidentified pesticides were recycled and stored according to regulations. The same is planned for 2005, but these actions do not solve the problem as a whole.

Unfortunately, the district is unable for the moment to use or destroy this dangerous toxic material. The optimum solution for this issue would be the storage or destruction by such entities that have the respective licenses for performing this kind of work.

The district will benefit the most from a method for destruction of this toxic material called "Mobile plasma complex for recycling of unfit and unidentified pesticides", which has been designed by the "Colorit" Institute of Dnepropetrovsk. However the mentioned method still has to be ecologically and sanitary-epidemiologically evaluated by experts.

A system of laboratory monitoring has been organized in the district which controls indications for pesticides and other chemical substances in objects of the environment, raw material supplies and food products.

The status of the atmospheric air and the water basins especially the water collectors and recreation sites is systematically checked for contents of toxic material.

Food products are systematically checked for safety indicators, including pesticides contents and chemical compounds. For this procedure highly sensitive GC and HPLC methods are used.

But a number of problems exist.

Currently to our regret, the laboratories of the state sanitary and epidemiological service lack modern indication instruments and express methods equipment, allowing a quick determination of the presence of toxic material.

With the development of chemical industry the moment has come to equip the laboratories of the sanitary and epidemiological services with modern instrumentarium and methods, capable quickly to establish small (insignificant) quantities of the different groups of poisoning material in the air, the water and the food products.

An issue with equal significance for the Dnepropetrovsk district is the provision of radiological safety for the population.

The sanitary and epidemiological service of the Dnepropetrovsk district also provides supervision for the observance of radiological safety rates during operation of ionizing radiation sources in public health, industry and other industrial branches. It carries out studies on food products, drinking water, construction material, gamma-survey of the territory.

Under special supervision is the 30 km zone of the Zaporozhsk NPS (nuclear power station).

Annually the radiological subdivisions of the sanitary and epidemiological service of Dnepropetrovsk district are conducting measurements for over 1000 different tests and for more than 2000 measurements of the power of the external gamma radiation dose on different objects and in open air.

Measurements of the dose load of about 500 individuals chosen amonprofessionals and random population, living in the 30 km zone are executed annually. Our specialists execute works on determining the affiliation of suspicious objects to the ionizing radiation source, on estimating the degree of danger, as well as organization of actions for its neutralization.

Radiological damages connected with loss, finding and appropriation of radioactive sources are not something unusual for the district. In 1999 such events have gained mass character in the Dnepropetrovsk district. As a rule such sources are used in different branches of production in radio-isotope instruments for the technological control of different processes. The sources "are packed" in special leaden blocks, which provide the required level of protection from the harmful effect of the ionizing radiation for the person. Considering the potential danger of the irradiation, the specialists of the sanitary and epidemio-logical service request higher caution from the workers providing the safety of these instruments.

However from 1999 till our days, a series of ravishment of ionizing radiation sources occurred on the territory of the district, and in the first place in the town of Krivoi Rog. From 1994 to 1998 only three such incidents were registered in the district, while in 1999 - five, three of which were connected to ravishment. It should be noted that all the stolen sources were of radionuclides caesium - 137.

The results of the analysis made by the sanitary and epidemiological service of the city of Krivoi Rog, revealed that the main reason that provoked the cases of ravishment was the formal attitude of the responsible persons towards their duties, for which they were punished in accordance with the sanitary legislation.

Fallen in the hands of criminals, which as a rule have a very remote notion on the danger of ionizing radiation, radioactive sources become a serious threat for the life and health of the terrorists themselves, as well as of casual persons. And the attempt to extract the ampoule with the radioactive source from its protection block is equal to expose oneself to radiation. Besides, in the event of destruction of the ampoule containing radioactive material, radioactive contamination of a large territory can occur. Such an incident happened already in Taromskoe suburb of Dnepropetrovsk.

Unpleasant"findings" often occur. Thus, in August 1999 during sorting scraps of the ferrous metal production of the ZAO "Ygprom" factory in the city of Krivoi Rog, a suspicious object was discovered labelled "Danger.Radioactivity". Further the object was identified as being a radio-isotope instrument for monitoring the degree of freezing of air-planes.

During the last years ionizing radiation radionuclide sources were repeatedly withdrawn from illegal circulation in Dnepropetrovsk,

Pavlograde, Dnepropetrovsk district, a small towns Carichanka and Megevaia.

In 2000 the law-enforcement organs arrested an inhabitant of small towns Carichanka when he was trying to sell a defensive block with a radionuclide source. After that, his house was inspected thoroughly and three similar blocks were discovered hidden at different places.

In September 2001 a citizen of Kiev was arrested in a suburb of Dnepropetrovsk, who was also trying to sell a defensive block with ionizing radiation source previously stolen in Doneck district.

After a year practically the same situation was repeated in Pavlograde city. There a Russian citizen tried to sell on the territory of Ukraine a radioactive source bought on the black market. In the course of the investigations it was discovered that the radioactive source was stolen from one of the enterprises in the Chelyabinsk district.

In all operations of the law-enforcement organs against illegal handling of the ionizing radiation sources, active participation took the specialists of the Dnepropetrovsk sanitary and epidemiological station.

Anxious to avoid possible income of nuclear waste with the scrap metal, the large metallurgical enterprises began introducing different systems for input radiation supervision. The most efficient from a number of operating systems is the automated "Cordon" in Krivoi rog metallurgical factory. Due to its installation, for a period from 2001 till now, it was not permitted for the smelting to be polluted by more than a group of ten radiation polluted batches. Amongst the radioactive waste most varied objects were discovered among which different nodes and spare parts polluted with caesium-37 from Chernobil, as well as natural radionuclide radium -226. There can be found radionuclide instruments also.

For the last 5 years under our participation 22 radiological incidents were liquidated.

According to our counts we took part in five successful joint operations against illegal possession of ionizing radiation sources with the law-enforcement organs.

An orderly system for interaction of the authorities on all levels including lifeguards, ecological, sanitary and epidemiological service, specialized from the radiological protection organizations was created in the Dnepropetrovsk district in order to counteract with a well-timed reaction to any information about suspicious for radiation objects. As a result of the coordinated activities of all interested services, now it takes less than one day from the incoming of the information on suspicious radiation object till its neutralization.

The facts about finding powder of unknown origin in postal envelopes shocked the world. In 2003-2004 on the basis of laboratory analysis for extremely dangerous infections, the regional sanitary and epidemiological station specialists investigated more than 44 mailings, containing unknown powdery material. Fortunately, according to the laboratory analysis, results in all the cases were negative.

In the majority of the countries, the necessary equipment and potential have been provided for to counteract the main affections from chemical incidents. These measures include not only the setting up of specialized organizations and institutions such as public health establishments or emergency committees and commissions, but also creation of reliable information sites and systems.

More efforts for the protection of the population from all types of terrorism are needed based on the already available systems and initiatives.

Chapter 3

BIOMONITORING OF EXPOSURE TO CHEMICAL WARFARE AGENTS

Noort, D., Van Der Schans, M.J., and Benschop, H.P.
Division of Chemical & Biological Protection, TNO Prins Maurits Laboratory, P.O. Box 45, 2280 AA Rijswijk, The Netherlands

Abstract: An overview is presented of the major methods that are presently available for biomonitoring of exposure to chemical warfare agents, i.e., nerve agents and sulfur mustard. These methods can be applied for a variety of purposes such as diagnosis and dosimetry of exposure of casualties, verification of nonadherence to the Chemical Weapon Convention, health surveillance, assessment of low level exposures (Gulf War Syndrome) and last but not least for forensic purposes in case of terrorist attacks with these agents.
This paper will focus on methods that are based on the analysis of long-lived protein adducts of CW agents which are detectable weeks or even months after exposure. Examples of real exposure incidents will be described.

Keyword: biomonitoring; diagnosis; DNA adducts; immunoassay; mass spectrometry; protein adducts, retrospective detectioin; sulfure mustard

1. INTRODUCTION

Methods to analyze chemical warfare agents (CWA) and their decomposition products in environmental samples were developed over the last decades (1). In contrast herewith, methods for such analyses in biological samples have only recently become available (2). Retrospective detection of exposure to CWA can be useful for various applications. It can be envisaged that rapid diagnostic methods can play a pivotal role in case of a terrorist attack with CWA. In the same context, confirmation of non-exposure of worried citizens is of utmost importance. Also, such methods can be used for forensic analyses in case of suspected terrorist activities ("chemical fingerprints"). It is self-evident that these methods will also be highly valuable from a military point of view, *e.g.,* to establish firmly to which chemicals casualties have been exposed to, which is a starting point for adequate medical treatment, or for health surveillance of workers in destruction facilities of chemical warfare agents. This

C. Dishovsky et al. (eds.), Medical Treatment of Intoxications and Decontamination of Chemical Agents in the Area of Terrorist Attack, 21–26.

presentation will focus on a number of specific methods currently available for verification of exposure to the most common CWA, *i.e.*, nerve agents and mustard agents.

There are basically four methods to diagnose an exposure to a nerve agent:

1. cholinesterase inhibition measurements
2. analysis of hydrolysis products, *e.g.*, alkyl methylphosphonic acids
3. analysis of generated phosphofluoridates after treatment of blood with fluoride ions ("fluoride reactivation")
4. mass spectrometric analysis of phosphylated peptides after enzymatic digestion of modified cholinesterase.

For mustards, there are three distinct methods to assess an exposure:

1. mass spectrometric analysis of low molecular weight urinary metabolites
2. analysis of DNA adducts by means of mass spectrometric or immunochemical methods.
3. mass spectrometric analysis of protein adducts, *e.g.*, to hemoglobin and albumin.

Presently available methods to diagnose and biomonitor exposure to anticholinesterases, e.g., nerve agents, rely mostly on measurement of residual enzyme activity of acetylcholinesterase (AChE) and butyrylcholinesterase (BuChE) in blood. More specific methods involve analysis of the intact poison or its degradation products in blood and/or urine. These approaches have serious drawbacks. Measurement of cholinesterase inhibition in blood does not identify the anticholinesterase and does not provide reliable evidence for exposure at inhibition levels less than 20 %. The intact poison and its degradation products can only be measured shortly after exposure. Moreover, the degradation products of pesticides may enter the body as such upon ingestion of food products containing these products.

In the case of sulfur mustard, analysis of low molecular weight urinary metabolites suffers from the same drawback as in the case of anticholinesterases, i.e., these products are rapidly excreted and provide therefore limited retrospectivity. Similarly, the in vivo lifetime of DNA adducts of sulfur mustard are less than those of protein adducts due to repair of DNA damage.

This paper will focus on methods that are based on the analysis of long-lived protein adducts, *i.e.*, on methods 3 and 4 for nerve agents and on method 3 for mustards.

2. NERVE AGENTS

Analysis of phosphofluoridates generated from phosphylated BuChE

In principle, organophosphate-inhibited BuChE in human plasma is a persistent and abundant source for biomonitoring of exposure to organophosphate anticholinesterases. Polhuijs et al. (3) developed a procedure for analysis of phosphylated BuChE in plasma or serum samples, which is based on reactivation of the phosphylated enzyme with fluoride ions: this converts the organophosphate moiety completely into the corresponding phosphofluoridate, which is subsequently isolated and quantitated. As in the case of analysis of hydrolysis products this approach identifies the organophosphate except for its leaving group. Moreover, the extent of the organophosphate poisoning can be determined in this way. Furthermore, based on the minimal concentrations of phosphofluoridate that can be analyzed in blood, it can be calculated that inhibition levels $\geq$ 0.1% of inactivated BuChE (*i.e.*, trace level exposure) should be quantifiable. Evidently, by analyzing the inhibited enzyme instead of the uninhibited enzyme, inhibitor levels can be measured that are several orders of magnitude lower than those based on residual AChE activity. The method is limited by spontaneous reactivation and ageing (i.e., loss of the alkyl moiety from the alkoxy moiety of the phosphyl group) of the phosphylated enzyme and by the natural life span of the enzyme. Application of this method to serum samples of the victims from the Tokyo subway attack and of the Matsumoto incident yielded sarin concentrations in the range of 0.2-4.1 ng/ml serum. Evidently, these victims had been exposed to an organophosphate with the structure $iPrO(CH_3)P(O)X$, presumably with X = F (sarin).

Depending on the structure of the nerve agent, a retrospectivity of 2-8 weeks was observed upon analysis of blood samples from rhesus monkeys which had been challenged with a sign-free dose of the nerve agent (4).

Mass spectrometric analysis of phosphylated peptides after enzymatic digestion of modified cholinesterase.

Recently, Noort et al developed a procedure that is based on straightforward isolation of adducted BuChE from plasma by means of affinity chromatography with a procainamide column, followed by pepsin digestion and LC/electrospray tandem MS analysis of a specific nonapeptide containing the phosphonylated active site serine-198 residue (5). This method surpasses the limitations of the fluoride-reactivation method, since it can also deal with dealkylated ("aged") phosphonylated BuChE. The method allowed the positive analysis of several serum samples of Japanese victims of the terrorist attack in the Tokyo subway in 1995. Furthermore, the method could be applied for detection of ChE modifications induced by, e.g., diethyl paraoxon and pyridostigmine bromide, illustrating the broad scope of this approach. This new approach will also allow biomonitoring/retrospective detection of exposure to several organophosphate pesticides and carbamates in one individual.

3. SULFUR MUSTARD

Mass spectrometric analysis of protein adducts

Sulfur mustard is a strong alkylating agent that reacts readily with nucleophiles under physiological conditions. The reaction products of sulfur mustard with these nucleophiles are all potential biological markers of human poisoning. Metabolites derived from an initial reaction with water and glutathione are excreted in urine (2). Adducts to DNA which may be present in various tissues and blood can conveniently be detected by using an immunochemical assay (6). This section will deal with ad-ducts to hemoglobin and albumin, since it is expected that they are persistent and will allow retrospective detection.

Upon incubation of human blood with sulfur mustard, it appears that 20-25% of the dose was covalently bound to hemoglobin. The most abundant adduct was the histidine adduct. In addition, the adducts to cysteine, gluta-mic and aspartic acid and to the N-terminal valine residues were detected (7). As a biological marker of poisoning, N-alkylated N-terminal valine has the advantage that it can be selectively cleaved from haemoglobin by a modified Edman procedure using pentafluorophenyl isothiocyanate as reagent. Analysis of the resultant pentafluorophenyl thiohydantoin, using negative ion GC-MS-MS after further derivatisation with heptafluorobutyric anhydride, provided a very sensitive method for the detection of the N-alkylated valine. *In vitro* exposure of human blood to ≥ 0.1 μM sulfur mustard and *in vivo* exposure of guinea pigs could be detected employing this method. Moreover, the adduct could be detected in samples from victims of accidental exposure to sulfur mustard and CW casualties (8). Recently, a standard operating procedure (SOP) for determination of the sulfur mustard adduct to the N-terminal valine in hemoglobin was developed (9). By using this SOP, it was found that the N-terminal valine adduct in globin of hairless guinea pigs and marmosets subsequent to i.v. administration of sulfur mustard (0.5 LD50) is persistent for at least 56 and 94 days, respectively.

Recently, sulfur mustard has been shown to alkylate a cysteine residue in human serum albumin (10). The site of alkylation was identified in a tryptic digest of albumin from blood exposed to [^{14}C]sulfur mustard. A sensitive method for its analysis was developed based on Pronase digestion of alkylated albumin to the tripeptide S-[2-[(hydroxyethyl)thio]ethyl-Cys-Pro-Phe, and detection using micro-LC-MS-MS. *In vitro* exposure of human blood to ≥ 10 nM sulfur mustard could be detected employing this method. The analytical procedure was successfully applied to albumin samples from Iranian casualties of the Iraq-Iran war.

4. CONCLUSIONS

- Adducts with macromolecules such as proteins offer long lived biological markers of exposure, possibly up to several months.
- Gas or liquid chromatography combined with tandem mass spectro-metry are the methods of choice for unequivocal identification at trace levels.
- Further improvements in retrospectivity and detection limits will hinge on future enhancements of sensitivity and resolution of electro-spray mass spectrometry instruments and of several hybrid configurations.

ACKNOWLEDGEMENTS

This presentation covers work that was funded by the US Army Medical Research and materiel Command, by the Bundesministrerium der Verteidigung, InSan I 3, Germany, and by the Directorate of Military Medical Service of the Ministry of Defence, The Netherlands.

REFERENCES

1. Chemical Weapons Convention Chemicals Analysis. Mesilaakso, M. (ed.), John Wiley, New York, 2005 (to be published)
2. Noort, D, Benschop, H.P., and Black, R.M. (2002). Biomonitoring of exposure to chemical warfare agents: a review. Toxicol. Appl. Pharmacol., 184, 116-126.
3. Polhuijs, M., Langenberg, J.P., and Benschop, H.P. (1997). New method for retrospective detection of exposure to organophosphorus anticholinesterases: application to alleged sarin vicitims of Japanese terrorists. Toxicol. Appl. Pharmacol., 146, 156-161.
4. Van der Schans, M.J., Polhuijs, M., Van Dijk, C., Degenhardt, C.E.A.M., Pleijsier, K., Langenberg, J.P., and Benschop, H.P. (2004). Retrospective detection of exposure to nerve agents: analysis of phosphofluoridates originating from fluoride-induced reactivation of phosphylated BuChE. Arch. Toxicol., 78, 508-524.
5. Fidder, A., Hulst, A.G., De Ruiter, R., Van Der Schans, M.J., Benschop, H.P., and Langenberg, J.P., (2002). Retrospective detection of exposure to organophosphorus anti-cholinesterases: mass spectrometric analysis of phosphylated human butyrylcholinesterase. Chem. Res. Toxicol., 15, 582-590.
6. Van der Schans, G.P., Scheffer, A.G., Mars-Groenendijk, R.H., Fidder, A., Benschop, H.P., and Baan, R.A. (1994). Immunochemical detection of adducts of sulfur mustard to DNA of calf thymus and human white blood cells. Chem. Res. Toxicol., 7, 408-413.
7. Noort, D., Hulst, A.G., Trap, H.C., De Jong, L.P.A., and Benschop, H.P. (1997). Synthesis and mass spectrometric identification of the major amino acid adducts formed between sulphur mustard and haemoglobin in human blood. Arch. Toxicol., 71, 171-178.
8. Benschop, H.P., Van der Schans, G.P., Noort, D., Fidder, A., Mars-Groenendijk, R.H., and De Jong, L.P.A. (1997). Verification of exposure to sulfur mustard in two casualties of the Iran-Iraq conflict. J. Anal. Toxicol., 21, 249-251

9. Noort, D., Fidder, A., Benschop, H.P., De Jong, L.P.A., and Smith, J.R., (2004). Procedure for monitoring exposure to sulfur mustard based on modified Edman degradation of globin. J Anal. Toxicol., 28, 311-315.
10. Noort, D., Fidder, A., Hulst, A.G., Wooffitt, A.R., Ash, D., and Barr, J.R. (2004). Retrospective detection of exposure to sulfur mustard: improvements on an assay for liquid chromatography-tandem mass spectrometry analysis of albumin-sulfur mustard adducts. J. Anal. Toxicol., 28, 333-338.

Chapter 4

NON-RULED MARKET ECONOMY AS A SOURCE OF CHEMICAL TERRORISM. AUTOMOTIVE FUEL: QUALITY AND ENVIRONMENTAL SAFETY

William Zadorsky
Ukrainian State Chemical-Technology University
8. Gagarin Ave., Dnepropetrovsk 49005 Ukrain

Abstract: The problems of the market economy influencing the state of environment in Ukraine are illustrated by the example of the Pri-dneprovie region. New examples resul, when ecologically dangerous enterprises and technologies are created in flagrant contradiction with the conception of sustainable development in order to please the requirements of the market in the region. Ecologically dangerous projects are realized, which on the basis of their effects can be named chemical terrorism without overstatement. Special attention was paid to the use in Ukraine of the "special" fuel on the base of wastes of coke factories of Ukraine, containing a carcinogen and benzenel in quantities, which exceed by orders of magnitude the doses accepted in other Europian countries and the rest of the world. Possibilities to decline the harmful influence of such fuel on the population of Ukraine in the market conditions are shown.

Keywords: chemical terrorism; cleaner production; fuel; liquid waste; market economy; rocket utilization; sustainable development

1. DISCUSSION

At long last, it was announced that, "based on the major ideas and principles declared at the Rio de Janeiro Conference of 1992, Ukraine deems desirable a shift to sustainable development that would ensure a balanced solution to social and economic tasks and to problems of leaving the environment and the potential of natural resources in good order for the current and future generations."

C. Dishovsky et al. (eds.), Medical Treatment of Intoxications and Decontamination of Chemical Agents in the Area of Terrorist Attack, 27–48.

As is well known, the concept of sustainable development includes three aspects, namely environmental, economic and social. Underestimating any of these facets will lead to a distortion in this equilateral triangle and to a deviation from the overall strategy of sustainability. This strategy can only be implemented when the three tasks are fulfilled simultaneously. The systems approach reveals strong interactions between the three factors of sustainable development. The sustainability will therefore be determined mainly by those parameters that affect at least two of the three factors.

The National Academy of Sciences of Ukraine suggested that an economic-environmental-social model be devised and employed for the purposes of the country's sustainable development. This is a very complex and time-consuming approach that may not be usable at this time of industrial restructuring, privatization and other involved processes occurring in a collapsed national economy. An alternative tactics is put forward, which is applicable at both national and regional level. Instead of mathematical modeling and optimization, it uses systems approach and decision theory techniques.

The environmental pollution in Ukraine has reached critical levels. Vehicles, power plants, steel mills and nonferrous metallurgy works are the major air polluters. Pesticides are responsible for much of the harm done to human health.

Water pollution continues on a large scale. The Dnieper region is the worst affected among its counterparts in Ukraine. Here, the situation is disastrous because of a combination of energy-intensive industries, thermal power generation, and intensive agriculture, further aggravated by Chernobyl.

There is an urgent need to find feasible ways that would stop the ensuing depopulation in Ukraine, such ways to ensure survival that would work before the sustainable development concept has been implemented. In these times of a deep economic crisis, the economic and environment-related issues must be attacked simultaneously, in line with one strategy for a cleaner economy. This would be a change from a policy of anthropogenic impact assessment to that of at-source abatement of pollution.

Such a program must be specific, realistic and not contradicting the idea of sustainability. It might be a program of cleaner industrial and agricultural production incorporating a systems principle of ascension from cleaner local units, mills, factories, and areas to a cleaner Ukraine to cleaner multinational regions.

What makes our approach rather different from the mainstream international cleaner production (CP) movement is the desire to abolish the dominating "black-box" techniques. Instead of regarding a production facility as no more than a given set of benign inputs and polluting outputs, we insist that one should seek the best ways to affect a prospective cleaner object within the "black box".

As the major principle of cleaner economy, the systems approach is taken that deals with perfecting any nature-technology system at the various hierarchic levels, from environmental pollution sources to consumers, and takes into account the interactions and mutual effects of all important components. This type of analysis will reveal relationships between the ways to improve processes and the challenges of risk management and nature conservation. The main task is therefore to harmonize the nature-technology relation and, ideally, to engineer high-performance systems featuring desired environmental characteristics at each hierarchic level, so that the favorable environmental background is not impaired and, where possible, even restored. Following are the basic assumptions underlying the cleaner economy concept for Ukraine:

- At this time of a deep economic crisis, the economic and environmental challenges must be met simultaneously, in keeping with one strategy of cleaner economy.
- A move towards a cleaner economy must focus not on consumption, but rather on perfecting those entities that are actual or potential polluters.
- The success of a cleaner economy policy will be largely determined by the availability of professionals well trained in the theory and practice of "economy clean-up" and environmental management.
- No cleaner economy will be possible without creating a civilized environmental market.

These strategic principles determine some tactical measures for pursuing them. Such measures are applicable to any industry and include:

- no waste due to improved selectivity,
- neutralizing wastes directly at the origin, rather than at the exit,
- flexible technologies,
- recycling materials and energy,
- conservation of resources,
- waste treatment, etc.

These tactics must be combined with certain design and process engineering techniques, such as

- providing a considerable excess of the least hazardous agent,
- minimizing dwell times,
- recirculation of materials and energy via closed loops,
- concurrent reactions and product separation,
- introduction of heterogeneous systems,
- adaptive processes and apparatuses,
- increasing throughputs,
- multifunctional environmental facilities, etc.

For a cleaner economy to be affordable an environmental market must be established and market mechanisms set to motion between all its interacting operators. Integration is necessary that would link researchers and developers of environmentally high technologies to designers of equipment to manufacturers to users. Professionals must be trained and further educated in the fields of industrial ecology and environmental management. Qualified consulting and assessing bodies are needed that would be capable of certifying environmental products, properly performing scientific, engineering and legal assessment, and winning public trust at the environmental market. Legislation is necessary that would give incentives to managers and entrepreneurs promoting cleaner production, ensure benefits to companies upgrading their production facilities to make them more environment-friendly, and stimulate development of an environmental market focused on high technologies, equipment, labor and services and having all proper attributes like competition, arbitration courts, commercial practices etc.

The latter paragraph is closely related to the idea of restructuring in the area of material production to be based on

- developing a socially oriented market economy that would guarantee a proper life standard for the population,
- cleaner production, minimizing environmental loads, material conservation, adoption of new types of activity grounded on environmentally safe technologies,
- making a more balanced economy by shifting from production of means of production to consumer goods, and
- environmental impact assessment and auditing for all economic projects.

The macroeconomic transformations rely on changes in the structure of production and consumption, mainly in industry. This necessitates:

- a more pronounced social orientation of industry to increase the relative importance of light and food-processing industries,
- an effective combination of industry branches keeping abreast of international requirements and meeting domestic needs,
- setting limits to raw material and semi-finished product industries,
- stepwise reduction of exports from primary and other material- and energy-intensive industries,
- increasing outputs of high-added-value products to facilitate effective utilization of domestic resources, and
- restructuring the production environment via introduction of recent scientific accomplishments, conservation of energy and other resources, implementation of waste-free and environment-friendly technologies, application of optimized power sources, waste treatment and utilization.

These points are common to the Draft Concept and to the cleaner production concept.

How could all that be funded? It would be ridiculous to suggest that comprehensive actions towards a cleaner economy might be supported by the miserable national budget. There is even less reason to believe that any down-to-earth effort to upgrade some specific facility and implement advanced technologies and state-of-the-art equipment might be funded from it. Furthermore, we can hardly rely on the assistance from the

West, and nobody here believes in assistance not backed by economic interest.

What is to be done to make cleaner production profitable, as is the case in the West? One source might be those fines that are imposed on violators of environmental laws in Ukraine. However, no mechanisms of channeling this money to environmental investments are available so far.

All over the world, the environmental market is replacing punitive methods of environmental management and those environment protection agencies that are not capable of coordinating and managing cleaner economy projects.

Worldwide, the taxation and payments for resources and emissions are devised in such a manner as to make it more profitable for the manufacturer to resolve the environmental issues in-plant, rather than to shift them off to the consumer area. A combination of sanctions with economic incentives for a cleaner production will make the latter not a recipient from, but rather a donor to the government budget.Yet even domination of a cleaner economy policy will not soon guarantee survival of the population under a deep economic crisis like the current one. The anthropogenic damage already caused to nature may prove too heavy and not lending itself to repair within the life span of one or even more generations.

An analysis of the man-production-environment system reveals that for survival of human beings the CP concept must be complemented by two more lines of action, namely

- adaptation of human body to life in adverse conditions, and
- utilization of life support systems.

The former approach implies development and implementation of biomedical and unconventional methods for prevention of ailments, adaptation and rehabilitation based on recent scientific findings and combined efforts of scientists, engineers, educators and managers under a degrading environment in Ukraine. These are the major lines to be pursued:

- setting up a tailored system of environmental education and training for the population in environmentally damaged areas, relying on the existing environmental education network and the media,
- research and development of adaptation promoters, immunogens and detoxicants, mostly of natural origin, processes and equipment for their manufacture and application practices,
- launching industrial production of adaptation promoters, immunogens and detoxicants, mainly from Ukraine's domestic starting materials, and
- research and development of existing and new non-medicinal methods of health building and adaptation to anthropogenic loads, including ways to reduce immune reactivity of and risks to people subjected to adverse working environments, residing in heavily polluted areas or dealing with ionizing radiation and other negative factors at work.

The latter concept implies providing local life support systems for unfriendly environments. By now, Ukrainian scientists and engineers have developed a variety of processes for potable water treatment by adsorption, electrochemical oxidation, electrocoagulation, electro-coprecipitation, electrodialysis, electrofloatation, floatation, membrane techniques etc. Each family must get small units for water purification, air cleaning and removal of hazardous substances from the food as soon as possible, for it may take decades to introduce cleaner production on a national scale. Here, we should follow the example of Western business people who bring with them to Ukraine devices enabling a safe existence in this unfriendly environment.

More specifically, environment professionals in Dnepropetrovsk have offered a number of local CP projects. One of them is concerned with treatment of ash of the local steam power plant. According to Canadian experts, 32 elements may be recovered from the ash in addition to the residual coal, making the business of ash treatment highly profitable.

There have been projects to produce building materials from the fly ash collected directly at precipitation filters. Moreover, this material attracted international entrepreneurs who wanted to export it to Spain, most probably for purposes of extraction of some rare earth metals. It is regrettable that no local business people took interest in the idea, especially when in Dnepropetrovsk area there are defense industry giants like Yuzhnyi Engineering Plant and Chemppri with their expertise in high technologies, including recovery of valuable metals and fabrication of appropriate sorbents and equipment.

The steam power plant should become another site for an exciting project enabling a 2-fold reduction in the degree of flue gas cleaning while cutting the electric power consumption by a factor of 2 to 3. The new process that applies pulsed voltage to the precipitation filters has been successfully introduced at several other plants in Ukraine.

These and other projects were included in the draft program of cleaner production for the Dnieper region. Each item in the program is backed with engineering and economic analyses. For many of the projects, international partners and prospective investors are sought that may gain profit by cleaning our environment.

Coal is one of the major fuels in Ukraine where its environmental impact is much higher than that of nuclear energy, provided that there are no accidents. Yet there is no alternative to coal, for natural gas can only cover the most urgent household and industrial needs, not to mention the country's indebtedness for its imports. The question of environment-friendly combustion of coal is therefore highly topical.

It is of special importance to the Lower Dnieper region where Pridneprovsk, Zaporizhia and Krivoi Rog steam power plants, Europe's biggest, are operated alongside with hundreds of smaller plants, cogeneration units and boilers. To abate pollution, one has to find its causes. With this in mind, a digest of recent materials found on the Internet is given in this issue. It has following sections:

The environmental pollution in transition economy countries has reached critical levels. Vehicles, power plants, steel mills and nonferrous metallurgy works are the major air polluters. Pesticides are responsible for much of the harm done to human health. Water pollution continues on a large scale. Very often the situation is disastrous because this country has a combination of energy-intensive industries, thermal power generation, and intensive agriculture, further aggravated by Chernobyl.

There is an urgent need to find feasible ways that would stop the ensuing depopulation in a lot of FSU countries, such ways to ensure survival that would work before the sustainable development concept has been implemented. In these times of a deep economic crisis, the economic and environment-related issues must be attacked simultaneously, in line with one strategy for a cleaner economy.

This would be a change from a policy of anthropogenic impact assessment to that of at-source abatement of pollution. Such a program must be specific, realistic and not contradicting the idea of sustainability. It might be a program of cleaner industrial and agricultural production incorporating a systems principle of ascension from cleaner local units, mills, factories, and areas to a cleaner country to cleaner multinational regions.

What makes this approach rather different from the mainstream international cleaner production (CP) movement is the desire to abolish the dominating "black-box" techniques. Instead of regarding a production facility as no more than a given set of benign inputs and polluting outputs, we insist that one should seek the best ways to affect a prospective cleaner object within the "black box".

As the major principle of cleaner economy, the systems approach is taken that deals with perfecting any nature-technology system at the various hierarchic levels, from environmental pollution sources to consumers, and takes into account the interactions and mutual effects of all important components. This type of analysis will reveal relationships between the ways to improve processes and the challenges of risk management and nature conservation. The main task is therefore to harmonize the nature-technology relation and, ideally, to engineer high-performance systems featuring desired environmental characteristics at each hierarchic level, so that the favorable environmental background is not impaired and, where possible, even restored.

Following are the basic assumptions underlying the cleaner economy concept for transition economy countries:

- At this time of a deep economic crisis, the economic and environmental challenges must be met simultaneously, in keeping with one strategy of cleaner economy.
- A move towards a cleaner economy must focus not on consumption, but rather on perfecting those entities that are actual or potential polluters.

- The success of a cleaner economy policy will be largely determined by the availability of professionals well trained in the theory and practice of "economy clean-up" and environmental management.
- No cleaner economy will be possible without creating a civilized environmental market.

The idea of restructuring in the area of material production to be based on:

- developing a socially oriented market economy that would guarantee a proper life standard for the population,
- cleaner production, minimizing environmental loads, material conservation, adoption of new types of activity grounded on environmentally safe technologies,
- making a more balanced economy by shifting from production of means of production to consumer goods, and
- environmental impact assessment and auditing for all economic projects.

The macroeconomic transformations rely on changes in the structure of production and consumption, mainly in industry. This necessitates:

- a more pronounced social orientation of industry to increase the relative importance of light and food-processing industries,
- an effective combination of industry branches keeping abreast of international requirements and meeting domestic needs,
- setting limits to raw material and semi-finished product industries,
- stepwise reduction of exports from primary and other material- and energy-intensive industries,
- increasing outputs of high-added-value products to facilitate effective utilization of domestic resources, and
- restructuring the production environment via introduction of recent scientific accomplishments, conservation of energy and other resources, implementation of waste-free and environment-friendly technologies, application of optimized power sources, waste treatment and utilization.

These points are common to the Sustainable Development Concept and to the cleaner production concept. How could all that be funded? It would be ridiculous to suggest that comprehensive actions towards a cleaner economy might be supported by the miserable national budget. There is even less reason to believe that any down-to-earth effort to upgrade some specific facility and implement advanced technologies and state-of-the-art equipment might be funded from it. Furthermore, we can hardly rely on the assistance from the West, and nobody here believes in assistance not backed by economic interest.

What is to be done to make cleaner production profitable, as is the case in the West? One source might be those fines that are imposed on violators of environmental laws in transition economy countries. However, no mechanisms of channeling this money to environmental investments are available so far.

All over the world, the environmental market is replacing punitive methods of environmental management and those environment protection agencies that are not capable of coordinating and managing cleaner economy projects.

Worldwide, the taxation and payments for resources and emissions are devised in such a manner as to make it more profitable for the manufacturer to resolve the environmental issues in-plant, rather than to shift them off to the consumer area. A combination of sanctions with economic incentives for a cleaner production will make the latter not a recipient from, but rather a donor to the government budget.

Yet even domination of a cleaner economy policy will not soon guarantee survival of the population under a deep economic crisis like the current one. The anthropogenic damage already caused to nature may prove too heavy and not lending itself to repair within the life span of one or even more generations.

An analysis of the man-production-environment system reveals that for survival of human beings the CP concept must be complemented by two more lines of action, namely adaptation of human body to life in adverse conditions, and utilization of life support systems.

The former approach implies development and implementation of biomedical and nonconventional methods for prevention of ailments, adaptation and rehabilitation based on recent scientific findings and combined efforts of scientists, engineers, educators and managers under a degrading environment in Ukraine. These are the major lines to be pursued:

- setting up a tailored system of environmental education and training for the population in environmentally damaged areas, relying on the existing environmental education network and the media,
- research and development of adaptation promoters, immunogens and detoxicants, mostly of natural origin, processes and equipment for their manufacture and application practices,
- launching industrial production of adaptation promoters, immunogens and detoxicants, mainly from domestic starting materials, and
- research and development of existing and new non-medicinal methods of health building and adaptation to anthropogenic loads, including ways to reduce immune reactivity of and risks to people subjected to adverse working environments, residing in heavily polluted areas or dealing with ionizing radiation and other negative factors at work.

The regional program of adaptation and rehabilitation of the population is developed for the first time in the Ukraine as for near Dniepr river region, and in a kind of the concept is offered for any of a technogenous intense region. Feature of the program in integration of efforts on joint activity in spheres of a science, engineering, training and management with the purpose of decision of problems of preventive maintenance, adaptation and rehabilitation of the population in conditions of worsened ecological conditions in the Ukraine.

Main directions of realization of the program:

- fulfillment of scientific researches on creation of adaptogenous, immunogenous (predominary a natural origin), techniques of their application, technologies and equipment for their manufacture;
- fulfillment of scientific researches on ordering known and creation of new nonmedicine methods healthbuilding and adaptation of the person to technogenous effects;
- creation of industrial manufacture of adaptogenous, immunogenous predominary based on Ukrainian raw sources;
- development and realization of the educational ecological programs for all categories of the population.

Main sections of the program:

1. Organization of a system ecological education and training of the population and realization of direct work in ecologically intense regions with use of a network of ecological training and mass media,
2. System engineering of diagnostic, adaptation, the increasing of the healthlevel and rehabilitation of the persons, received infringements of a health as a result of technogenous effects.
3. Fulfillment of scientific researches on ordering known and creation of new non-medicine methods of healthbuilding and adaptation of the person to technogenous effects, in particular, development of ways of immunity increase and reduction of the factors of risk at the persons, subjected to effect of harmful industrial factors, 4. Fulfillment of scientific researches on creation adaptogenous, immunogenous and other medicines for resistance against acting of harmful substances (predominary a natural origin), techniques of their application, technologies and equipment for manufacture. Survey of natural raw sources of adaptogenous, immuno-genous and development of industrial technologies and creation of its manufactures.
6. Development and introduction of new food products on the basis use natural byoaddings with adaptogenous and immunogenous properties, ensuring preventive maintenance of diseases ecological ethyology.
7. Development and organization of manufacture of an ecological engineering and surviving means.

With arrival in Ukraine market economy it is possible to mark that many ecological problems both at the regional level, and at level of our country, were intensifyed. We have not while systematized analytical data about the change of concentrations of harmful matters in the water, air, soil, but of indirect indexes put a number on one's guard.

But every year the population of Ukraine diminishes approximately on half-million persons, more anxious data about child's death rate and morbidity. In Dnepropetrovsk to the end reasons are not found out of epidemic growth in the last years of quantity of pulmonary diseases. It is possible to consider as results of biotesting periodically intensifyed tendency to disappearance of populyatsii of sparrows in the center of city, Lately the complete disappearance is noticed practically in the town of cockroaches. But, speak, they are led even in the nuclear reactor...

1.1. Processes of the Restructuring, Privatization, Military Conversion in the countries with transitional economy with account of market conditions.

During Restructuring, Privatization, Military Conversion there is inconsistencies between requests of maintenance of sustainable development and problem of deriving of maximum profits. In the countries with transitional economy unfortunately there is slacking of state and municipal management and while there is not market mechanisms of managing of SMEs activity. Recently new program on Market Mechanisms and Incentives for Environmental Management (URL: http://es.epa.gov/ncer/rfa/02marketmech.html) as part of U.S. Environmental Protection Agency (EPA) and of its Science to Achieve Results (STAR) is announced. They are seeking applications for research leading to improved theoretical and/or empirical analyses of the feasibility and effectiveness of market mechanisms and economic incentives (MM&I) as substitutes for, or complements to, traditional environmental management programs. The terms 'market mechanisms' and 'incentives' refer to approaches that rely on economic incentives, market forces, or financial mechanisms to encourage regulated entities to reduce emissions, discharges and waste generation, or generally improve environmental performance. EPA is interested in supporting research on practical applications of MM&I approaches related to its mission, i.e., addressing environmental quality and human health.

1.2. Chemical Terrorism on local, national, and international levels (Source: A FOA Briefing Book on Chemical Weapons)

There have always been fears that terrorists might be tempted to acquire and use weapons of mass destruction. The world received a shocking reminder of the potential impact of terrorist use of chemical weapons when the Aum-Shinrikyo sect used the nerve gas Sarin to attack civilian targets in Japan during 1994 and 1995. It is fortunate that, to-date, few other incidents of this nature have occurred. In the wake of the horrific terrorist attacks in New York on September 11th, 2001, fears of terrorist use of chemical and biological weapons were rekindled, and fueled by public release of information suggesting that the terrorists had considered using crop dusting aircraft in their plans.

There is much that local, national, and international authorities can, and should, be doing to counter the threat of chemical terrorism. For an overview of specific subject of the protection against chemical weapons, it is necessary to pay attention on:

- Decontamination

- Detection
- Protective Equipment
- Collective protection

Protection against Chemical Weapons. There are four main cornerstones in the protection against chemical weapons, all of which are largely dependent upon each other to provide optimum effect. These four are:

- physical protection: body protection, respiratory protection, collective protection,
- medical protection: pretreatment, therapy,
- detection: alarm, monitoring, verification, identification, all-clear,
- decontamination: individual decontamination, equipment decontamination.

An overview of chemicals defined as chemical weapons. Main Groups:

- Nerve Agents
- Mustard Agents
- Hydrogen Cyanide
- Arsines
- Psychotomimetic Agents
- Toxins
- Potential CW Agents

Today, only a few of these are considered of interest owing to a number of demands that must be placed on a substance if it is to be of use as a CW agent.

- A presumptive agent must not only be highly toxic but also "suitably highly toxic" so that it is not too difficult to handle.
- The substance must be capable of being stored for long periods in containers without degradation and without corroding the packaging material.
- It must be relatively resistant to atmospheric water and oxygen so that it does not lose effect when dispersed.
- It must also withstand the heat developed when dispersed.

CW agents can be classified in many different ways. There are, for example, volatile substances, which mainly contaminate the air, or persistent substances, which are involatile and therefore mainly cover surfaces.

All decontamination is based on one or more of the following principles:

- to destroy CW agents by chemically modifying them (destruction),
- to physically remove CW agents by absorption, washing or evaporation,
- to physically screen-off the CW agent so that it causes no damage.

Most CW agents can be destroyed by means of suitable chemicals. Some chemicals are effective against practically all types of substances. However, such chemicals may be unsuitable for use in certain conditions since they corrode, etch or erode the surface. Sodium hydroxide dissolved in organic solvent breaks down most substances but should not be used in decontaminating skin other than in extreme emergencies when alternative means are not available.

CW agents can be washed and rinsed away, dried up, sucked up by absorbent substances, or removed by heat treatment. Water, with or without additives of detergents, soda, soap, etc., can be used, as well as organic solvents such as fuel, paraffin and carburettor spirit. Emulsified solvents in water can be used to dissolve and wash-off CW agents from various contaminated surfaces.

Unfortunately, all these methods can not used in the case of global chemical terrirism of some modern "market" technologies are used in Ukraine. I would like to name only some of them on the example of our Pridneprovie Region:

1. During last several years one by one there are already 4 manufactures of the lead accumulators " Ista " adjoining directly to two inhabited files of city. I shall remind, that the building of new ecologically dangerous manufactures is forbidden in the technogenic overloaded city Dnepropetrovsk approximately 20 years ago.
2. Recently in the center of city Dnepropetrovsk it was utilization a lot of rockets "SS-20" (Program of USA – Ukraine) and we had the pollution in the air a lot of very dangerous and toxic geptil.
3. Using of the most cheap and therefore dirty and dangerous sorts of coal and liquid fuel (mazut) by our Pridneprovie Heat Electric Station,
4. And at last it is the using of the "special" Automotive fuel on the base of liquid wastes of coke factories of Ukraine, containing a carcinogen and drug benzol in quantities that are exceeding in other countries of Europe and world in tens times.

1.3. Automotive fuel on the base of liquid wastes of coke factories of Ukraine as the chemical terrorism factor.

Recently Cleaner Production Center organized an Environmental Symposium on Transportation and Environment. In recent years, the number of cars in Dnepropetrovsk increased dramatically, predominantly owing to import of used cars, and exceeded 250,000 in 2000. Since the city has no by-pass highways, tens of thousands of cars pass its center daily. They run on cheap fuels containing massive admixtures of aromatic compounds like benzene, toluene and xylene. The fuels produced by coke-and-byproduct plants in Zaporozhe and Dneprodzerzhinsk contain 16 times as much benzene as does US gasoline (1 to 2 %). In Kharkov, 100 gas-filling stations sell gasoline of unacceptable quality.

The Security Service of Ukraine in Kharkov blocked an attempt of two local gasoline stations to sell adulterated gasoline. The hazardous fuel included the authentic A-76 grade heavily mixed with benzene, toluene and xylene to imitate high-octane grades A-93 and A-95.

A good news, too. It is regrettable, however, that in Dnepropetrovsk very many, if not all, cars have to use fuels adulterated with these and other hazardous compounds. The practices of admixing them to gasoline have been carried on in this area for several years, in spite of numerous protests by concerned environmentalists in the media. This badly aggravates the environmental situation, already disastrous in the entire region.

Xylene, benzene, styrene and toluene total to 40 to 60 per cent in most of the fuels, as reported by Dnepropetrovsk Sanitary Monitoring Station.

The highway police, the air quality monitoring services or the meteorological stations do not measure the levels of these compounds. Although no reports on the amounts of the above aromatics in the exhaust gases are available, experts assert that complete combustion of these compounds in an engine is impossible. The combustion products therefore pollute the air to a degree that makes the common CO and CO_2 pleasant air fresheners. No catalytic converter will help. Rather, it will be poisoned prematurely.

Several measures may help remedy the situation with fuels:
(a) to introduce new governmental specifications for the fuels to limit the aromatic contents,
(b) to impose restrictions on sales of adulterated fuels based on more stringent control of fuel quality,
(c) to carry out certification of bodies having a right to sell automotive fuel,
(d) to put heavy environmental taxes on low quality fuel to be paid by both the producer and the distributor,
(e) to get involved the governmental standardization bodies and consumer societies, and
(f) to use state-of-the-art additives enabling a more complete fuel combustion and therefore improved environmental performance of vehicles.
Only the latter option is workable at present, because the country is lacking oil. Following are some additives already in use in Ukraine.

General-Purpose Fuel Modifiers T-4 and T-6 developed by A. Ozeryanskii in Kiev, Ukraine. Non-toxic agents that reduce pollutant discharges 1.5 to 3-fold and offer 8 to 15 % savings in fuel, depending on engine wear and fuel type, up to 30 % reduction in carbon monoxide, and 50 % less smoke, while increasing the engine power by 10 to 28 %. They are very simple in use, so that a 1.2-ml ampoule is just emptied into a fuel can. One liter of T-4 is enough to modify 18,500 l gasoline or 5,600 l diesel fuel. Its cost is 2 to 3 % of the fuel cost.

Oil Modifier MP-8 of the same inventor reduces oil losses by a factor of 1.6, extends oil service life 2- to 3-fold and cuts pollutant discharges by at least 30 %, while enhancing engine power by up to 10 %. Also, this non-toxic additive reduces wear and varnish formation on engine parts and improves detergency and sealing. Its cost is about 40 % that of domestic motor oil.

Some other developments may also prove useful for making combustion engines cleaner.

Clean Air Valve developed by Ted Switen, USA. Tests on carburetor engines in Ukraine showed 2- to 4-fold reductions in the exhaust levels and fuel economy improvement by 1 %. At its prospective price of about $50, the device will pay back after a 15,000 to 20,000 km run. A modification intended for diesel engines will be run in in 1998.

Ukrainian Catalytic Converter developed in Kiev. Being similar to its Western counterparts in performance, the device will be less expensive. Ukraine committed itself to the European Union that all its vehicles would be equipped with catalytic converters by 2003.

Binary Fuel Device developed and produced on a full scale in Ukraine. Reduces hazardous exhausts by a factor of 2 to 3 through the use of a low-octane gasoline plus propane-butene. The breakeven is predicted at 20,000 km.

Yet another approach is to influence the combustion process itself by ionization, pulsed electromagnetic fields, glow discharge etc. The Institute of Energy at Dnepropetrovsk State University developed a plasma-activated sparking plug, currently in preparation for a full-scale production.

Prof. V.N. Nabivach (Ukrainian State University of Chemical Engineering, Dnepropetrovsk). The levels of polycyclic aromatic compounds (PACs) in exhaust gases deserve special attention, for they often exceed maximum permissible concentrations 2 to 2.5-fold. A carcinogenic constituent like benzopyrene is million times as hazardous as CO and 50,000 times as harmful as NOx, an important difference being that its action is prolonged. The contribution of automotive engine exhausts to overall urban air pollution in Ukraine is approaching 30 %, a trend towards the 50 % in the West. In the absence of standards that would limit aromatics in automotive fuels, the coke and byproduct plants like those in Dneprodzerzhinsk, Zaporozhe and Avdeevka are free to produce fuels containing 50 to 70 % PACs, compared to "only" 30 to 40 % in the products of oil processing companies. PACs are fairly stable and tend to be adsorbed on building walls, trees etc., thus posing a permanent threat to public health. Another headache is dioxin, an extremely hazardous pollutant discovered in automotive exhaust gases in 1980s.

A.G. Khandryga (Manager, Industrial Product Certification Dept., Dnepropetrovsk Center for Standardization and Metrology). We do certify the fuel that is imported whenever its quality is acceptable. What happens next, when it is unloaded into storage tanks and further on, is beyond our control. Only the Agency for Consumer Protection can request a certificate of quality directly at a gas-filling station. It may then see tens of certificates issued at different places. The law does not prohibit a distributor to apply to a certification center other than the local one. Furthermore, one check-up revealed that about 80 % of the certificates were forged.

Prof. L.M. Pritykin (State Academy of Civil Engineering and Architecture of Dnieper Region). A rapid method to determine quality of gasoline is needed. The existing instruments generally measure octane number. This is not sufficient, for even with an acceptable octane number the exhaust gases may be highly toxic and the engine life short. Our effort was focused on a simple device for checking on gasoline grade. Our method uses 1 to 2 drops of fuel and takes 1 or 2 seconds to determine whether the gasoline really is of the labeled grade. The hand-held instrument does not need utility power and can be used in the field. It is fabricated in Ukraine and priced at $150 to $200. When the seller's grade does not correspond to the instrument reading, one should refer to more specific techniques for determination of the fuel fractional composition, octane number, chemical group composition etc. It is highly desirable to set up a center for fuel check-up on an absolutely independent basis. Such a center should be unbiased, self-standing and open to everybody. It might be run by the city executive committee but never by the commercial companies dealing in fuel.

V.M. Golushko (Deputy Manager, Agency for Customer Protection). We have a right to check the quality of petroleum products since late 1997. We have inspected about 30 gas-filling stations and in 80 % of cases had to block the sales because of lacking certificates or poor quality. The problem of mixing various grades does exist but lends itself to control. Another potential trouble is the location of gas-filling stations that are often situated near the Dnieper River. No precautions against leaks are made.

As Prof. V.M. Nabivach, Dr. V.A. Gerasimenko informed about problem of automotive fuel: quality and environmental safety, motor vehicle pollution is ever increasing, a great many hazardous compounds being discharged near the ground not only on highways but also in housing areas. Of all the ingredients of automotive fuel, aromatic hydrocarbons are the most toxic, benzene being the worst for its carcinogenic properties. Toluene and xylene have narcotic action. Fuel composition is a major factor that determines which compounds are present in the exhaust gases. Over 500 organic compounds have been reported in exhaust gases, including the initial ingredients and also polycyclic aromatics, of which 3,4-benzopyrene is the nastiest. The level of this carcinogen in exhaust gases increases with increased initial concentration of aromatics in the fuel. Other data suggest that lowering emissions of benzopyrene and fuel soot, its carrier, is the quickest way to a healthier urban atmosphere.

The gasolines used in industrialized countries display high octane numbers while having no more than 43 % aromatics and up to 2 % benzene. As regards Ukrainian products, the respective values may be at 62 % and 4.6 % for gasolines produced by oil refineries and as high as 77 % and 26 % for products of coke plants.

In order to achieve dramatic reductions in hazardous emissions to air in urban areas, it is therefore recommended that the domestic standards for gasoline limit benzene to 2 % and total aromatics to 45 %.

As informed Dr. A.I. Korableva from Institute for Environmental Management and Ecology under the National Academy of Sciences of Ukraine in the report "Environmental impact of automobile transport by example of Dnepropetrovsk", Dnepropetrovsk with its annual discharge of air pollutants of 177,000 t (as of 1996) is among the worst affected cities in Ukraine. In these, the automobile transport was found to be responsible for at least 30 % of the total emissions which are 15 times the maximum permissible level. Aside from the dust, chemical, photochemical and noise pollution, there is the aspect of street washout of automobile-related pollutants into the River Dnieper. The measured annual receipts of lead, particulates and petroleum derivatives via rainwater and thaw water to the river are 0.45, 80,000+ and 1.8+ t respectively. The actual levels of petroleum derivatives in storm water sometimes were 206 times the maximum permissible concentration (MPC) for the fishery basins. At 34 km downstream from the city, the estimated levels of petroleum derivatives and particulates are 61 and 10.8 times the respective MPCs. The airborne lead is mainly accumulated in the soil of housing areas.

And, at last, Prof. E.A. Derkachev, Dr. L.B. Ogir, Dr. A.A. Shevchenko, A.P. Shtepa, V.V. Sotnikov, V.I. Sviridov, G.P. Isaeva from Dnepropetrovsk State Medical Academy it their report "Hygienic assessment of automobile transport impact on the environmental situation and public health in Dnepropetrovsk" informed, that:

1. Soil, being one of the most stable elements of the environment, may serve as an indicator of long-time pollution with specific agents like heavy metals.
2. The distribution of lead in the soil in Dnepropetrovsk suggests that the pollutant is mostly supplied by automobile transport via exhaust gases.
3. Lead pollution level is predictive of a number of diseases in the affected population.
4. Reliable correlations were found between lead concentrations in the soil and the incidence rates in the population of Dnepropetrovsk. In the adults, the pollution levels were associated with complications in pregnancy and delivery, diseases of urogenital, nervous and bone-and-muscular system, blood problems and tumors. In the children, the blood, blood forming organs, respiratory system and the muscles and bones were most frequently affected.

5. Comprehensive measures for abatement of automobile-related pollution in Dnepropetrovsk are needed.

The Symposium adopted the following resolution that deems it necessary:

1) to introduce new governmental specifications for automotive fuels with a view to limiting the aromatic contents in harmony with international standards, to effect more stringent control of fuel quality and to impose restrictions on sales of adulterated fuels;

2) to implement state-of-the-art computerized systems for assessment and prediction of transportation effects on environment;

3) to carry out certification of bodies having a right to sell automotive fuels and to put a heavy environmental tax on low quality fuel to be paid by both the producer and the distributor;

4) to use state-of-the-art additives enabling a more nearly complete fuel combustion and therefore improved environmental performance of vehicles;

5) to highlight that accidents involving hazardous freights, particularly massive at railroads, are a challenge. According to the incomplete data available, hundreds of these occur at NIS railroads, so that one-time discharges of hazardous substances pose considerable threats to people and environment. The environmental monitoring of transportation is not up to the mark, advanced air purification and water treatment processes are not employed on a sufficient level, the waste management is inadequate, and the work on cleaner transportation has just started;

6) to consider as particularly urgent the problem of prevention and elimination of consequences of accidents in transportation of hazardous freights in the territory of Ukraine, and to develop proper techniques and facilities for the relevant transportation bodies;

7) to use the experience in environmental information monitoring gained by the Laboratory for Environment Protection at the Pridneprovskaya Railroad and to have the Ecotrans information retrieval system cover the entire railroad network of Ukraine;

8) to start R&D on waste management for the transportation companies;

9) to implement ideas and practices of constructive ecology, like closed-circuit water systems, on a broader scale;

10) to deem it necessary to gear Dnepropetrovsk Center for Standardization and the Consumer society to attack on the overall problem;

11) in view of the positive feedback from environmentalists in the region, to launch regular publication of Constructive Ecology and Business journal in 1999 and further on. To appeal to management of major industrial and transportation companies for support; and

12) to inform Dnepropetrovsk City Council, the Ministry of Transportation and the Ukrainian Railroad Company about the Symposium output and resolutions.

Besides I would like to inform you about the current situation with automotive fuels in Russia, mainly in Moscow, and in Kazakhstan

(Almaty) from the viewpoints of engine operation, fuel production, commercial trends and government regulation. International experience in these fields is also covered. Considering the quality of fuels available in Russia (and Ukraine), ways to achieve a more nearly complete combustion are to be preferred to reliance on catalytic converters. A number of additives are covered, including the following ones produced in Russia:

Molilat: Lubricity additive containing molybdenum disulfide for improved resistance to wear, pitting and fretting. Reduces noise, increases maximum speed and engine life and cuts fuel consumption.

Molilat-2M: Metal-depositing additive for recovering compression in the cylinders and pressure in the lubrication system of worn engines. Improves power performance, saves fuel and oil, reduces noise and smoking of the exhaust. Guaranteed compatibility with all mineral oils.

Udav: Lubricity additive applicable for new and old engines alike and compatible with any mineral lubricant. Contains ultrafine diamond powder for special friction. Reduces noise, wear and smoking while improving power and extending engine life.

Aspekt-Modifikator-U: Single-use agent which produces a thin porous film capable of retaining lubricant on the surface of engine and transmission components. Reduces noise, wear, varnish formation and fuel consumption and improves detergency.

Aspekt-Modifikator: Versatile fuel additive that neutralizes water, prevents corrosion and ice formation in the fuel system while improving engine environmental performance.

RiMet: Lubricity additive for motor and transmission oils. Contains ultrafine particles of a special alloy. Reduces wear and smoking, cuts fuel and oil consumption and improves compression.

Turbo-Oktan 115: A small-pack version of Feterol for increasing the octane number directly in the fuel tank. The agent eliminates detonation, enhances power and extends engine life while its excess of oxygen improves combustion and thus reduces pollution.

Forum: Wear-preventive additive containing polytetrafluoroethylene particles 0.4 micrometers in diameter. Dramatically reduces noise and oil loss.

As for Ukraine, it has Renoks, an antidetonation modifier developed by Prof. Yu. Merezhko of the Ukrainian State University of Chemical Engineering in Dnepropetrovsk. It offers a 1.5 to 2-fold reduction in pollutant emissions and an 8 % cut in fuel consumption at the cost of 1.0 to 1.5 % of fuel price. Also available are Remol-2 and Remol-2A produced by Remol in Odessa. These oil additives can be recommended for both carburetor and diesel engines after 40,000 km run to recover and stabilize compression, save up to 8 % fuel, protect engine components and reduce hazardous substances in exhaust gases.

The multipurpose fuel modifiers T-4 and T-6 produced by ADIOZ company of Kiev were covered in the previous issue.

Mazut fuel emulsification. The Dnepropetrovsk national university for a number of years deals with the complex problems of power system connected to economy of fuel and protection of an environment. In this sense the direction connected to use water-fuel emulsion is perspective. Within the framework of this direction emulsification systems for liquid fuel are developed in view of individual conditions of the customer. The systems developed by their emulsification fuel are successfully maintained on some enterprises of Ukraine.

The known data are put in a basis of development on an intensification of process of burning and reduction of toxic emissions at burning in furnace (the chamber of combustion) water-fuel emulsion. The comparative data on burning waterless and emulsified fuel show, that emulsified fuel burns down much faster waterless, thus burning water-black oil emulsion in furnaces provides economy of 10-15 % of black oil in comparison with burning of clean fuel. Besides one of the factors determining efficiency of use water – fuel – emulsion (WFE) in котельно-топочных processes, the opportunity on their basis is to solve a number of environmental problems. WFE burning reduces a pollution emissions of NOx in gas (approximately on 50 %), approximately in 3-4 times reduces emission of carbon black adjournment, reduces emissions of benzapirene in 2-3 times, etc.

The greatest economic benefit and simultaneous decrease of gas emissions provides addition in fuel of 10-15 % of water, and the greatest ecological effect regarding recycling the waters polluted with organic products is realized at a level of a water phase up to 50 %. The physical features of burning distinguishing WFE as new fuel in comparison with known, are defined substantially by imperfection existing sprayer devices. The last practically are not capable to provide dispersion of liquid fuel on a level of dispersiveness of less than 100 microns. At the same time the drops of emulsion with such size, created by sprayer devices contains some thousand micro- drops of water. To this in a high-temperature zone of furnace chambers the drop of emulsion blows up and occurs secondary dispersion fuel. This effect is especially important at burning coal pitch and heavy extra-heavy oils used in metallurgy. As a result of such microexplosions in furnace there are centers of turbulent pulsations and the number of elementary drops of fuel due to what the torch increases in volume increases and in regular more intervals fills the furnace chamber that results in alignment of a temperature field of furnace with reduction of local maximal temperatures and increase in average temperature in furnace; to increase of luminosity of a torch due to increase in a surface of radiation; to essential reduction underburning of the fuel; allows to lower quantity of blown air and to reduce connected with it heat loss. Simultaneously in a torch occur the catalytic reactions conducting to reduction of harmful gas emissions. The opportunity of reduction of quantity of blown air at burning water-fuel emulsions is rather important, as according to skilled given efficiencies of the boiler unit at reduction of factor of surplus of air on 0,1 increases for 1 %.

And at last I would like to inform you about our proposals at Systems approach to cleaner transportation using. The transportation-highway complex is a major air polluter, with the motor vehicles accounting for about 70 %, the rolling stock 25 %, the aircraft 2 %, the road-building machines 1.4 % and the vessels 1 % of the total transportation-related pollution. The morbidity in urban areas was found to correlate with the size and intensity of operation of the vehicle fleet. In rural areas, carcinogens originating from vehicle exhaust gases tend to accumulate in plants grown near highways.

The general issue of cleaner transport may be approached using the approximation of a closed system including man, nature and transport in which various feedbacks and feedforwards exist between all subsystems. Following specific environmental problems may further be recognized that relate to: running gear, engines,fuels,cargo carriage,cargo reloading, passenger compartments, infrastructure, and accidents.

The systems approach applied here enables treatment of hierarchic levels for each specific type of transport.

A quantitative characteristic of environmental safety and an algorithm of environmental safeguarding is suggested that may help find and select the best engineering solutions.

The algorithm includes the following steps:

1. Processing the initial information to determine the appropriate hierarchic levels and especially those that control the pollution.
2. Selecting methods to influence the system.

A major aspect of environmental safeguarding is the abolishment of «bulk» neutralization of pollutants in a mixed fluid flow. It must be replaced by local action taken, wherever possible, on a component-by-component basis as close to the origin as may be, and preferably inside the source. This is just the opposite of currently prevailing systems where the entire spectrum of pollutants are collected and neutralized and/or recycled. Moreover, the local treatment performed close to the source is less expensive, if one takes into account all relevant costs. Since vehicles mostly rely on combustion for their energy, the safeguarding methods will include:

- minimizing residence time and providing an excess of one of the reagents, resulting in reduced formation of side products,
- recuperation or looping of the matter and energy flows enabling a more nearly ideal combustion and reduced rates of side processes,
- heterogenization to suppress formation of side products by the removal of the target product from the reaction zone at the instant of its formation, and adaptive processes and hardware for more reliable operation due to improved flexibility, helping reduce discharges during idle running and acceleration.

3. Engineering and economic analysis of the available safeguarding approaches and methods.
4. Selecting the economically and environmentally optimal safeguarding alternative.

5. Making request for proposal to design the safeguarding system adopted.

6. Preparing a business plan to carry out the safeguarding project.

Any comprehensive project to process wastes and side products should not be developed before the completion of the safeguarding designing for the transportation system.

The most effective direction for ecologization of the fuel burning process is the use of water- fuel emulsions with the soluble in water catalytic non- corroding additions and ultrasonic mixing by the sharp superheated steam. At development of our suggestions we came from that fact, that exhaust gases of combustion engines contain a generous amount of the organic and inorganic compounds, distributed between the gas phase and phase, being microparticles, 90% from which have sizes less than 1 mk. These particles are adsorbed on its surface of hundred chemicals, including mutagennie and carcinogenic matters. Mass of these particles is very small, and therefore its are inhalable dust. A gas phase contains, including benzapiren and oxides of nitrogen, which are predecessors of acid rain. These exhausts cause at man the allergic reactions and can lead to different pulmonary and cardiovascular diseases, including the cancer.

Decreasing of the oxides of nitrogen and microparticles emissions, traditionally includes the use of catalytic converters. These methods are very difficult and require the large capital investments, that limits their commercial applicability. Addition of water to the fuel, multiplies mass of the air-fuel mixture, entered in time unit, resulting in the improved dispersion and mixing.The improvement of interfusion causes more high speed of freed of heat and growth of pressure, that results in formation of more high maximal pressure in the combustion chamber of engine. Water also lowers a spades temperature of burning fuels, that in same queue conduces to the diminished formation of oxides of nitrogen.

Reduction of the formation of oxides of nitrogen proportionally to maintenance of water in fuel and relies on many factors. Addition of 8-25% water allows to decrease of CO emission to 40-70%, soot - to 40-60%, heavy hydrocarbons (including benzapiren) – to 70%.

Unlike the known approaches we develop the use of water- fuel emulsions with the soluble in water catalytic non- corroding additions and ultrasonic mixing by the sharp superheated steam for reduction of the oxides of nitrogen and soot emission in the combustion engines. Our know how is also the use of the electro-activated water for emulgation. The first results testify the technical and economic expedience of the chosen direction.

Chapter 5

ROLE OF THE CHEMICAL WEAPONS CONVENTION IN COMBATING CHEMICAL TERRORISM

Jiri Matousek
Masaryk University, Faculty of Sci., EU Res. Ctre of Excellence for Environmental Chemistry and Ecotoxicology, Kamenice 126/3, CZ-625 00 Brno, Czech Republic

Abstract: The Chemical Weapons Convention (CWC) is shortly characterised stressing its main principles, inter alia the General Purpose Criterion. Status of its implementation as of December 2004 shows the main data obligatory declared by already 167 States Parties and main achievements in destruction of Chemical Weapons (CW) stockpiles and destruction / conversion of CW production facilities and the verification efforts. The Organisation for the Prohibition of the Chemical Weapons (OPCW) is briefly presented, results of the 1st Review Conference and future problems of the CWC are analysed. Special emphasis is laid on threats and benefits of the scientific and technological development and potential misuse of toxic chemicals for terrorist purposes. Role of the CWC especially of respective national implementation measures in combating chemical terrorism is analysed stressing the OPCW's expertise including its developed system of assistance and protection under the CWC and enforcement by all countries of the CWC's requirement to make the development, production, stockpiling, transfers and use illegal for anyone.

Key words: Chemical Weapons (CW), CW Convention (CWC), CW Production Facilities (CWPF), CW destruction, CWPF destruction & conversion, CW non-production, Organisation for the Prohibition of Chemical Weapons (OPCW), First Review Conference, Chemical terrorism .

1. INTRODUCTION

The Convention on the Prohibition of the Development, Production, Stockpiling and Use of Chemical Weapons and on Their Destruction, shortly described as Convention on general and comprehensive prohibition of chemical weapons, or Chemical Weapons Convention, abbreviated as CWC, was adopted in 1992 after complex negotiations on the basis of The Conference on Disarmament (and previous multilateral negotiating fora in Geneva), lasting nearly a quarter of a century mainly due to the worldwide spread of chemical industry and relatively easy

C. Dishovsky et al. (eds.), Medical Treatment of Intoxications and Decontamination of Chemical Agents in the Area of Terrorist Attack, 49–57.

possibility of clandestine production of chemical warfare agents in militarily relevant quantities. This is reflected in very careful definitions and criteria, defining purposes not prohibited by the Convention and mainly very complex and sophisticated verification system. The CWC is a very impressive and the best elaborated disarmament document, totally outlawing one important and very dangerous kind of weapon of mass annihilation, committing the States Parties (SP) to destruction of the chemical weapons (CW) stockpiles and of production facilities (CWPF). Ten years after opening for signature and six years after entry into force (EIF), The First Conference of the SPs reviewing operations of the CWC was convened in The Hague (2003), stating generally good acceptance by the international community, showing positive results of implementation of the CWC provisions, and defining the course for the future. CWC and its implementation is one important instrument in combating chemical terrorism, even if not universal due to the variety of its forms and material sources as it is explained below.

2. CHEMICAL WEAPONS CONVENTION – BASIC FACTS

The Chemical Weapons Convention (CWC) opened for signature in Paris, on January 13, 1993 and entered into force on April 29, 1997. Its complexity is reflected in almost 200 pages of text, containing Preamble and 24 Articles and three Annexes: On Chemicals (6 p), On Implementation & Verification (105 p), and On Protection of Confidential Information (5 p) [2].

To the main pillars of the CWC belong:

- Verified destruction of chemical weapons (CW) and CW production facilities (CWPF), i.e. disarmament,
- Verified non-production of CW, i.e. non-proliferation,
- Assistance and Protection,
- International Cooperation.

The genie of this Convention lies *inter alia* mainly in the mood of defining the scope of the prohibition. The CWC is rather purpose than compound oriented. This means that it is nothing like the list of prohibited compounds as some less informed people could expect. The CWC´s leading principle, which is often reported as General Purpose Criterion (GPC) is contained in the wording of Article II, para 1, defining the purposes of the CWC among "Chemical Weapons":

Article II DEFINITIONS AND CRITERIA

For the purposes of this Convention:

1. "Chemical Weapons" means the following, together or separately:

(a) Toxic chemicals and their precursors, except where intended for purposes not prohibited under this Convention, as long as the types and quantities are consistent with such purposes

(b) Munitions and devices, specifically designed to cause.....

(c) Any equipment specifically designed for use.....

Under purposes non prohibited by this Convention according to Article II para 2 (a) – (d) are understood:

"industrial, agricultural, research, medical, pharmaceutical or other peaceful purposes, protective purposes, namely those, directly related to protection against toxic chemicals, military purposes not connected with the use of CW and not dependent on the use of toxic properties of chemicals as a method of warfare as well as law enforcement including domestic riot control ".

Toxic chemicals are further defined in Article II para 2 as meaning:

"Any chemical which through its chemical action on life processes can cause death, temporary incapacitation or permanent harm to humans and animals. This includes all such chemicals, regardless of their origin or of their method of production and regardless of whether they are produced in facilities, in munitions or elsewhere".

From this explanation, quoting relevant articles of the CWC is evident, consistently with the mentioned GPC that the Convention:

a) is nothing like a list of prohibited compounds,

b) covers *any* toxic chemical intended to be used for chemical warfare (and therefore developed, produced and stockpiled), pursuant to Article II, para 1 (a) and para 2, including even those not yet synthesised. This means that the CWC is open-ended and the prohibition covers any scientific and technological development.

The CWC lists (only as the verification instrument) the most important toxic chemicals and their precursors, endangering the CWC (known in the time of drafting the CWC) within three Schedules, constituted according to the risk the chemicals pose for the Convention. Schedule 1 contains super-toxic lethal chemicals and key precursors that have no peaceful uses, Schedule 2 contains less dangerous toxic chemicals and precursors produced in small quantities, and Schedule 3 lists toxic industrial chemicals (that were in the former history used for chemical warfare) and precursors produced on mass scale. A frequent misunderstanding occurs considering the Schedules as the lists of "prohibited compounds" although it is clearly stated in the CWC that *"Schedules do not constitute a definition of CW"*. The open-ended prohibition however does not mean that toxic chemicals (including other than those contained in Schedules) cannot appear on battlefields being used by non States Parties or less possibly by SPs breaching the CWC or more possibly by the terrorist groups. That is why the scientific and technological development is to be very cautiously watched, international verification measures extended, national authorities and operation systems established, and respective legislation adopted in order to enable prevention and adequate response in real time (repression, protection, rescue and recovery) in cases of emergency.

3. STATUS OF IMPLEMENTATION OF THE CHEMICAL WEAPONS CONVENTION

(If it is not otherwise stated, the data on implementation are reported as of December, 2004).

- At present, there are altogether 167 States Parties to the Convention. Important is the membership of all P-5 members of UN Security Council and vast majority of states with declarable CWC facilities.
- Five SPs (Russia, USA, India, South Korea and Libya) declared possession of CW.
- Among SPs, there are 12 possessors of former (after 1946) CW production facilities (CWPFs), i.e. Russia, USA, India, South Korea, France, UK, China, Iran, Japan, Bosnia & Hercegovina, Serbia & Montenegro (the last two countries declaring the same facility), and Libya.
- The CWC implementation & verification regime now covers 90 % of world´s population, but what is more important, 98 % of world´s chemical industry.

Reviewing the figure on the number of SPs, it is also important to note that there are 16 signatory states that have not yet ratified (*inter alia* Israel) and altogether 11 countries that have not even signed. In addition to not very important states it is necessary to note the DPR of Korea and the neighbours of Israel (Egypt, Iraq, Lebanon and Syria) bounding their signature on the Israel´s withdrawal from its nuclear weapons programme. With regard to the hot region of the Near East, a very significant breakthrough has been made by the accession of Libya in 2004.

Assessing the universality of the CWC (by the way one of the requirement of the First Review Conference), one can come to interesting results comparing this requirement with the status of other principal agreements on weapons of mass destruction (WMD) as can be demonstrated by table 1. It seems that one could be satisfied with relatively high number of SPs, seven years after EIF in comparison with other presented arms-control / disarmament agreements. Nevertheless, for the prevention of any use of CW, it is necessary to reach higher number of SPs mainly because most of the above mentioned important non-SPs concentrated in Near and Middle East and on Korean peninsula are supposed nearly certainly to be possessors of CW (not to speak on possession of other kinds of WMD like in the case of Israel).

Table 1. Universality: CWC compared with other main agreements on WMD

Treaty	Entry into force	SPs	other signatories	non-signatories
NPT	1970	187	---	7
BTWC	1975	151	16	27
CWC	1997	167	16	11

The worldwide status of CWC implementation is witnessed by another important data:
- 155 initial declarations (on possession / non-possession of CW) were obtained from SPs (September 2004),
- 134 national authorities were established in the SPs (September 2004),
- 97 national legislations on implementing the CWC adopted (September 2004).

Especially the latter two numbers are still quite insufficient taking into consideration the tasks of such governmental office in the national implementation measures starting with the respective legislation and then supervision of the domestic chemical industry and any cooperative activities with the Organisation for the Prohibition of Chemical Weapons (OPCW).

The most important data from the declarations of SPs (see table 2) show the worldwide problems with possession, storage, former production of CW as well as with the spread of chemical industry as the point of outcome not only for the destruction of CW at present and in the near future but for monitoring the non-production of CW in chemical industry in future.

Table 2. Important data from the declarations by the SPs (as of September 30, 2004)

Subject	Declaring SPs	Declared sites
CW storage facilities (CWSFs)	6	35
CW destruction facilities (CWDFs)	6	42
CW Production Facilities (CWPFs)	12	64[a]
Abandoned CW	3	15
Old CW	11	43
Schedule 1 Chemicals Production	21	27
Schedule 2 Chemicals Production	35	429
Schedule 3 Chemicals Production	33	506
Discrete Organic Chemicals Production	72	4763

[a] Of the 64 reported former CWPFs, 47 have been certified as already destroyed & converted

The total number of declared sites (5947) which are to be regularly or randomly inspected shows the high burden of expected verification activities. At this stage of implementation, the verification activities have been obviously concentrated on storage and destruction, and in industry on facilities producing scheduled chemicals. The number of inspections totalled 1936 at 784 sites in 67 countries till December 10, 2004.

At present, the most important activity in implementing the CWC is the CW destruction:

- Declared chemical agents ~ 71.373 thousand tonnes
  Destroyed (November 30, 2004) ~ 10.351 thousand tonnes
- Declared munitions (containers) ~ 8.671 M items

Destroyed (November 30, 2004) ~ 2.145 M items

As expected, the destruction is proceeding asymmetrically, meeting domestic financial and technological problems with construction of destruction facilities including problems of acceptance by population in the respective regions. It is therefore expected, that the scheduled 10 years term for total CW destruction according to the CWC will not be managed and the allowed exception to extend the destruction period for another 5 years has been already asked for, by the Russian Federation and also by the US.

4. ORGANISATION FOR THE PROHIBITION OF CHEMICAL WEAPONS – OPCW

Pursuant to the CWC, after its signature, the Preparatory Commission was founded and after EIF the Organisation for the Prohibition of Chemical Weapons (OPCW) was established with HQ in The Hague / The Netherlands (Johan de Wittlaan 32, 2517 JR Den Haag). See also: http//:www.opcw.org.

The Organisation consists of three main elements:

a) Conference of the State Parties (all SPs, meets regularly once a year), present
Chair: H.E. Amb. Mr Krzysztof Paturej (Poland).

b) Executive Council (41 members distributed among the SPs on a regional, rotating base for 2 years term, meets regularly 4 times a year), present Chair:
H.E. Amb. Mr José Antonio Arróspide del Busto (Peru).

c) Technical Secretariat (507 staff members from 66 nations, of them about 200 inspectors), Director General: H.E. Amb. Mr Rogelio Pfirter (Argentina),

- Subsidiary bodies: Scientific Advisory Board (20 independent experts), Confidentiality Commission, Advisory Board on Administrative and Financial Matters.

5. THE FIRST REVIEW CONFERENCE

The character and tasks for the Conference were determined as follows:

- Review operations of the Convention,
- Take account of scientific and technological development,
- Lessons learned and recommendation for future implementation,
- Not an amendment (revision) conference.

The attendance represented then (May 2003) 113 SPs, 2 signatory states (Haiti, Israel), 2 non-signatory states (Libya, Angola), 5 International Organisations (ESA, ICRC, PCA, CTBTO, UNIDIR), 22 NGOs and 6 Industry Associations.

The Conference did not result in a radical change of direction for the OPCW or substantive decisions on crucial, still outstanding issues (e.g. so called "non-lethal" agents, riot control agents, "law enforcement", nil declarations in respect of OCPFs and like. The problems of scientific and technological development were tackled only very generally.

A number of priorities have, however been clearly recognised. To those priorities belong:

- Universality of the Convention,
- National implementation measures,
- International Cooperation and Assistance,
- Verification regime for the chemical industry
- Optimisation of verification measures
- Scientific and technological development and
- Functioning of the OPCW.

The detailed explanation goes beyond the framework of this paper. For further information see the adopted documents. This is in the first line the Political Declaration containing 23 paras [3] and the main written result, i.e. the Review Document with 134 paras [4]. Except many statements, mostly only general, the programme did not go too deep into the problems of impact of scientific & technological development on the CWC that are obviously connected with its implementation in future. This problem was analysed in the document prepared by the OPCW Scientific Advisory Board introduced in the Note by the Director General [5]. It is generally expected that this will mainly influence future activities of the OPCW, considering the crucial importance of the impact of scientific and technological development on the CWC mainly after current CW stockpiles of the SPs have been ultimately destroyed.

6. ROLE OF THE CWC IN COMBATING CHEMICAL TERRORISM

It is clear that the main reason for adopting the CWC was not combating chemical terrorism but preventing any material preparations and use of CW in wars and armed conflicts in the first line by state actors. Secondly, it is necessary to note that the problem of chemical terrorism is broader, encompassing not only misuse of CWs but also any non-weaponised toxic compounds. To the third form and material source of chemical terrorism belong intended strikes against industrial and social infrastructures of modern civilised societies followed by releases of toxic, liquefied and inflammable chemicals with secondary effects on humans and biota [6]. Nevertheless, the CWC is an important instrument in combating the most dangerous forms of chemical terrorism. The effort of OPCW and especially of national authorities in the States Parties ensure that chemicals produced for peaceful purposes are not misused, provide some guarantees that terrorists will not be able to acquire or make their own CWs. That is why the desired universality of the CWC and respective national implementation measures including adoption of relevant domestic legislation are of utmost importance. The enforcement by all countries of the CWC´s requirement to make the development, production, stockpiling, transfers and use of CWs illegal for anyone means that terrorists could be put on trial for violating the Convention.

The OPCW´s expertise and knowledge of CWs including its developed system of assistance and protection under the CWC (Article X) as a reflection of international solidarity and co-operation are being put to use to prevent and respond to chemical terrorism and thus considerably diminish its potential consequences. It can be added that pursuant to the recent UN Security Council Resolution No 1540, all nations are obliged to take actions ensuring that non-State actors cannot develop, produce, use or trade CWs according to the terms of CWC.

7. CONCLUSIONS

Operations of the Chemical Weapons Convention are proceeding satisfactory judging according to the status of its implementation by 167 States Parties and verification by the Organisation for Prohibition of the Chemical Weapons in The Hague seven years after entry into force. The First Review Conference stressed the importance of achieving worldwide universality in order to totally eliminate the heredity of past chemical arsenals once forever, prevent threats and utilise benefits of the scientific and technological development for the CWC implementation in the foreseeable future. The Chemical Weapons Convention, even if originally not dealing with chemical terrorism executed by non-State actors explicitly, plays undoubtedly important role in combating chemical terrorism due to expertise of OPCW, responsibility of national implementation measures, assistance and protection under the CWC and international co-operative activities in prevention, repression, protection, rescue and recovery, motivated to a very broad extent by the CWC and its implementation.

REFERENCES

1. The author, Professor Jiří Matoušek, PhD, DSc., Dipl. Eng., was working nearly 40 years in R&D for Chemical Corps, Medical Services and Civil Protection, *inter alia* as Head of the Czechoslovak NBC Defence R&D Establishment (now Military Technical Institute of Protection, Brno) and a member of the Czechoslovak Delegation to the Conference on Disarmament in Geneva. He is currently chairman of the OPCW Scientific Advisory Board.
2. UN (1993): Convention on the Prohibition of the Development, Production, Stockpiling and Use of Chemical Weapons and on Their Destruction. UN, New York.
3. OPCW (2003): Political Declaration. www.opcw.org/cwrevcon/doc/NAT/FRCPolitical declaration.html. OPCW, The Hague.
4. OPCW (2003): Review document. www.opcw.org/html/global/wgrc/2k3/rc1revdoc.html. OPCW, The Hague.

5. OPCW DG (2003): Note by the Director General: Report of the SAB on Developments in Science and Technology, OPCW, Conference of the SPs, RC-1/DG.2, 23.04.2003. OPCW, The Hague.
6. Matousek J.: Forms and material sources of CBRN-Terrorism. Symposium on nuclear, chemical and biological threats – A crisis management challenge, Jyväskylä 2003. Proceedings, University of Jyväskylä Research report No 98, 2003, pp 2-6.

Chapter 6

CHEMICAL INCIDENT SIMULATOR: A NEW APPROACH FOR DERIVING PASSIVE DEFENCE REQUIREMENTS

M.J.G. Linders, C.A. van Beest, P. Brasser, L.F.G. Geers, G. van 't Hof, R.A. Rumley-van Gurp, R.P. Sterkenburg, S.C. van Swieten, H.W. Zappey, and A.R.T. Hin
TNO – Defence, Security and Safety, The Netherlands

Abstract: Passive defence requirements depend strongly on the 'perceived' threat and the scenarios considered 'likely'. Traditionally, such aspects were addressed more or less subjectively by defence experts. The 'Chemical Incident Simulator' (CIS) - a chain of linked simulation models - simulates the dispersion of chemical warfare agents, detector responses, the effects of protective equipment, and the human toxicological responses for large numbers of scenarios considered realistic given a certain threat. The calculations start by simulating 'the agent release and transport' in an incident scenario which results in concentration-time profiles at the locations of detectors and personnel. The detector module generates an alarm-time profile when a chemical agent is detected. Based on the concentration-time and alarm-time profiles the skin and respiratory protection models calculate exposure profiles, using given protective equipment characteristics. Finally, the toxic effects module translates the exposure profiles into casualty probabilities for the personnel. Operational behaviour, like changes in 'Dress State' and typical reaction times are taken into account in the simulation.

All input parameters, scenario definitions and results are stored in a database for easy access and retrieval. Analysis of individual scenario results and statistical analysis over all scenarios (or any subset) is possible. Typical individual scenario results are deposition, dosage and casualty level on the attacked target. Typical statistical analysis results are dosage and deposition threat spectra, and casualty spectra. The casualty levels and spectra can be obtained for various health effect levels (eye effect, incapacitation, lethal) and protection levels (no protection, suit only, mask only, mask and suit, collective protection). This model thus largely eliminates the subjectivity involved in scenario studies, protective and detector equipment procurement.

Keywords: casualties; chemical incidents; detection; medical countermesures; passive defence; protective clothing; respiratory protection; risk analisis

C. Dishovsky et al. (eds.), Medical Treatment of Intoxications and Decontamination of Chemical Agents in the Area of Terrorist Attack, 59–70.

1. INTRODUCTION

Political and military tools are available to counter the threat of biological and chemical (BC) weapons and agents. Non-proliferation and disarmament are political and technical instruments that are more or less successful. In this respect, the Chemical Weapons Convention is a prime example of successful political measures in the chemical arena. This Convention prohibits the production, storage, handling and use of chemical weapons and all declared stockpiles have to be destroyed no later than April 2007. In spite of the successes of non-proliferation and disarmament, military tools are indispensable as well. Military defence measures, both active and passive, should aim at neutralising an imminent BC-threat. In the past decades, NATO's strong airpower has almost completely eliminated BC challenges delivered by military aircraft. Currently, a lot of attention is focused on missile defence to annihilate ballistic missiles as means to deliver chemical and biological weapons.

Traditionally, passive defence has been the preferred way to counter the BC-threat, however. Passive defence encompasses the whole array of measures that are available to the soldier: detection and identification, physical protection, medical countermeasures and decontamination.

Events during the last decade have shown that weapons of mass destruction may be used by terrorists as well. As a result of possible amateuristic and opportunistic behaviour of terrorists, the threat spectrum has been stretched accordingly. The scale of the attacks as well as their locations (urban, indoors) have changed as well. Furthermore, NATO has changed its doctrine for the planning and execution of operations. Operations other than warfare are executed frequently, e.g. crisis response operations and peace enforcing, peace keeping and humanitarian operations. Finally, identifying passive defence measures has become a matter of concern for civil authorities as well.

Modelling and simulation are increasingly important instruments and these approaches have an enormous impact in the area of passive defence. Traditionally, the assessment of the chemical threat using the concept of challenge levels has been the primary focus. As the threat picture is changing, assessment of biological threats as well as threats exerted by (industrial) toxic compounds (including releases other than attack) have become important issues as well.

Threat assessment has been used as a starting point to define the requirements for a passive defence system. In the past, such requirements were determined on a more or less *ad hoc* basis. At the TNO Prins Maurits Laboratory we have started a scenario-based systemic approach to model the complete chain of passive defence measures, in order to derive challenge levels and casualty levels. This enables us to study the effects of passive defence requirements upon these levels, thus improving the selection process.

2. CHEMICAL INCIDENT SIMULATOR

The Chemical Incident Simulator, CIS, simulates events that encompass the passive defence against chemical warfare agents. The model starts in 'release, transport and dispersion' mode, where the agent release in an incident scenario is simulated. In this mode the model generates concentration-time exposure profiles for the detectors, mask, suit, filters and people present in the scenario. In addition, challenge levels to the whole target are calculated. In the next step the model is in detection mode; as soon as the release of a chemical agent is detected, an alarm is generated. These detection alarms and the exposure profiles are input for the next mode, where the skin and respiratory protection models are triggered. These models calculate the amount of protection offered by the protective material. This results in exposure profiles for lung, eye and skin to liquid, vapour and aerosols. In the final mode the toxic-effects model translates the exposure profiles into casualty probabilities for the personnel.

Scenarios, i.e. attacks or incidents, are needed as input for the model. Over the years an extensive number of scenarios have been collected. For easy retrieval of scenarios a database has been built. The scenario takes into account all relevant factors necessary to calculate challenge levels, i.e. target data, weapon characteristics, chemical agent properties and meteorological effects. Furthermore, different NBC-alert states (Mission Oriented Protective Posture – MOPP) can be selected. These states range from 'low', meaning no protective clothing or mask is worn, to 'high', meaning the soldier is completely protected. Each alert status is characterized by time intervals that define how long it takes before the mask and suit are worn, thus offering their respective protection. The resulting challenge levels, dosage fields and deposition fields, are stored in the database as well.

In the future, the system will ideally provide an analysis tool to support planning and decision making. The system will support stand-alone operation in an analytical mode as well as interfacing with an integrated warning and reporting network to provide real-time analysis capability. Ideally, the system should also be capable of interfacing with other models that simulate the effects of blast, fragmentation, fire, nuclear events and combinations thereof. Finally, it should be noted that the systemic approach could also play a role in defining research policies, as it will be capable of pointing out relative weaknesses in the passive defence system which need improvement.

2.1. Release, transport and diffusion

For release, transport and diffusion the simulation program RAP2000 is used. The engine of RAP2000 consists of a series of models that predict physical quantities like concentration and surface deposition as function of time and location, given a chemical or biological release scenario. Together these models are indicated as RAP, the Risk Analysis Package. The latest version RAP2000 is capable of: 1) easily creating many (variations of) scenarios; 2) simulating the individual scenarios; 3) storing and updating the scenarios and results in a database; and 4) analysing the scenario results. The results of model calculations can be analysed individually (per scenario) or statistically (using any desired subset of the scenarios database).

A major premise of RAP is that every chemical or biological attack, including line shaped spray releases, can be split up in single sources. A single source is defined as a cloud of vapour and liquid drops with a three dimensional Gaussian mass distribution. A single source itself is split up in an initial vapour puff and a number of puffs containing droplets with the same size. These initial puffs have the same geometry as their ancestor single source. Within RAP movement and dispersion of puffs is handled analytically for the major part.

RAP simulations are thus based on a concept with a clear hierarchy. The lowest level is the puff. Vapour and droplet puffs make up a single source and finally one or more single sources define a scenario (see Figure 1). Single sources are allowed to appear at different times and locations in the scenario. After generation of the single source and its puffs, drops start to fall and evaporate, generating additional vapour. At the same time the drops are transported by atmospheric motion. Finally, drops might reach the surface. The corresponding puffs then become "lying" puffs and start to generate secondary vapour. For all these processes RAP contains physical models. Apart from the release, transport and diffusion models RAP also contains a module that handles blast and fragment dispersion associated with artillery shells filled with chemical warfare agents.

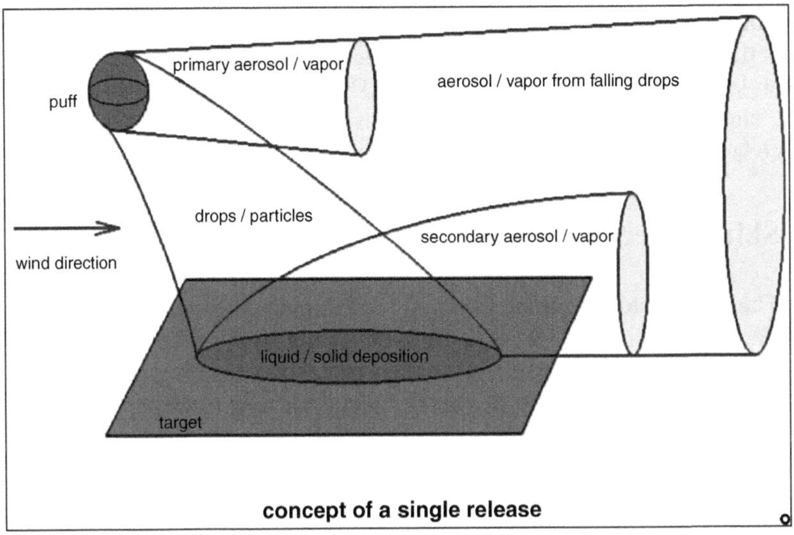

Figure 1. Release, evaporation and dispersion models of liquid and vapour in RAP.

2.2. Detector model

The detector model is capable of simulating both vapour and liquid detection systems. So far, about a dozen different detector systems are available. The detector model input signal (see 2.1) consists of: i) a time concentration profile, ii) the identity of the chemical warfare agent (HD, GB, VX or L) and iii) the relative air humidity (RH < 80% or RH > 80%).

For vapour detection there are three aspects that are modelled: sensitivity, response time, and regeneration. The sensitivity determines at which concentration level the detector will respond. The theoretical detector output (alarm or no alarm) is calculated by comparing the input data (concentration, relative humidity) with empirical detector display outputs, obtained during controlled exposure laboratory experiments. The response time determines how long it takes before the detector actually shows the response and it depends on the concentration level. The regeneration time determines how long it takes, after a positive detection, before the detector can do a new measurement.

The liquid detector model simulates the behaviour of detection papers, which are in operational use by the Dutch and many other Defence forces. It simulates whether or not a paper will show a visible coloration, depending on deposition density and droplet sizes.

The theoretical detector display outputs are corrected for operational detector procedures and residual contamination. The output signal is used in both the respiratory and skin protection models to increase the protection level when the signal switches to "alarm", and to decrease the protection level when it switches to "clear". The complete process of retrieval of input signal, performance of calculations and the generation of output signal needs only a few seconds.

2.3. Skin protection model

The skin protection model, or suit model, calculates the concentration of warfare agents, which penetrates the NBC-clothing. This concentration is determined by the state of protection: when protective clothing is worn, this concentration is determined by the breakthrough of agents through the suit, and when no protective clothing is worn, the skin concentration equals the concentration in the surroundings of the soldier. The vapour is adsorbed on the carbon, which is present in the NBC-protective clothing. Due to the characteristics of the clothing material, there will always be a certain breakthrough concentration of vapour. The breakthrough concentration is calculated by the model on the basis of the type of NBC-protective clothing material, the outside concentration, the temperature, the wind speed, the time of exposure, the type of vapour etc. Next to vapour contamination, the suit model also includes a basic liquid drop model. When liquid drops hit the clothing material, they will start to evaporate, and this vapour can also penetrate the clothing. Currently, liquid breakthrough is not yet covered by the model because of the complexity of this matter. Several models for liquid breakthrough are being tested at the moment. The liquid drop model is being extended to take into account the effects of wicking and wetting of the material by liquid. All of the aforementioned processes together result in a breakthrough concentration as a function of time.

2.4. Respiratory protection model

The respiratory protection model, or mask model, consists of two parts: a carbon filter model and a mask leakage model. The carbon filter model predicts the vapour breakthrough through the filter as a function of time. The model is valid for the adsorption of a vast number of physisorbed organic contaminants. Climatic aspects like temperature and humidity are important parameters in this respect. Different – but constant – scenario temperatures are possible, provided that all parameters are known or can be estimated at the temperature of interest. Humidity has not yet been incorporated. During actual use of a gas mask the flow through the filter is not constant. A breathing cycle will be incorporated in the model by applying a sine wave pattern, which closely resembles the actual breathing pattern. Furthermore, the breathing volume must be included as well, as it depends on the status of the soldier: breathing is more intense during work than in a state of rest.

The leakage model is deduced from protection factor measurements of people wearing gas masks in the field. The leakage is expressed as a distribution of protection factors as it varies quite a lot over a population of people. The final vapour concentration that a soldier inhales and to which the eyes are exposed, is a fraction-based mean of the breakthrough through the filter and of the leakage at the sides of the mask.

2.5. Toxic effects model

The toxic effects model uses concentration-time profiles from the respiratory and skin protection models as input to estimate casualty probabilities. Two approaches are available: a simple linear dose-effect model as incorporated in RAP and a more elaborate non-linear response model, based on the Toxic Load approach. The latter provides a better description of toxic effects for agents that show significant deviations of simple Haber's law behaviour (i.e. toxic responses only depend on the concentration-time product and not on each quantity separately).

Toxic effects of expositions are calculated for a variety of exposures and effect combinations, assuming a probabilistic dose-effect relationship. Lethal and incapacitating responses (e.g. respiratory effects, topical skin effects or incapacitating eye effects) of varying degrees of severity are addressed. The model also distinguishes between effects resulting from vapour exposure and from exposures to liquid droplets. These primary effect probabilities are subsequently combined to afford overall casualty probabilities for lethality, severe incapacitation and incapacitation due to topical eye effects.

The model is also capable of addressing the effects of various medical countermeasure protocols upon casualty flows, e.g. when chemoprophylaxis and/or specific therapeuticals for intoxications are available. At present, this aspect is not parameterized properly and therefore is available only as a prototype to demonstrate proof of principle.

The toxic effects model relies on many consensus parameters that describe toxicological and pharmacological effects. These consensus parameters are the result of an in-house review of available toxicological data. Other parameters (e.g. those derived from NATO study groups) may, however, be used when necessary or desired.

Future developments will include the development of consensus parameters that describe realistic medical countermeasure effects. An important new feature will consist in the assessment of casualty flow onset times, by calculating the dose-build-up and the resulting development of casualty probabilities in time.

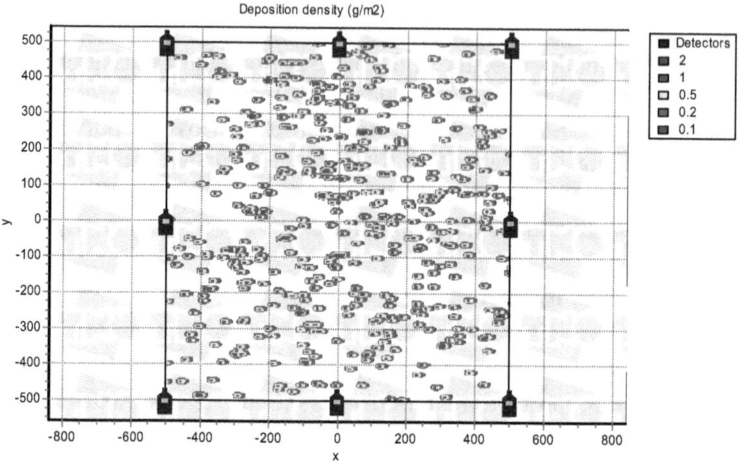

Figure 2. Liquid deposition density of sarin on the target area.

3. RESULT AND DISCUSSION

The Chemical Incident Simulator simulates the dispersion of chemical warfare agents, detector responses, the effects of protective equipment, and the human toxicological responses for large numbers of scenarios. The possibilities and potentials offered by the Chemical Incident Simulator are illustrated best with an example.

The calculations start by defining the scenarios: incident properties such as target, terrain, climate, weapons, agent etc.; personnel deployments – type of protection available (mask, suit); detector deployments – single detector, array of detectors, location; and NBC-alert state. Subsequently, the 'agent release and transport' in the scenario is calculated, which results in concentration-time profiles at the locations of detectors and personnel. To illustrate the concept, the impact of 740 small submunitions from a TBM, releasing the nerve agent sarin, is used. Figure 2 depicts the liquid deposition density as a result of the attack.

The locations of the detectors, an array of eight, are shown in Figure 2 as well. The detector model generates an alarm-time profile when a chemical agent is detected. As the concentration exceeds a given threshold, the detector status changes from "clear" to "alarm". After the concentration drops below the threshold, the status reverses to "clear" again. In case of multiple detectors in the field, an overall alarm profile is generated, which is clearly shown in Figure 3. Eight alarm profiles plus the overall profile are shown (scaled between 0 and 1), each corresponding to the local concentration profile. As soon as any of the individual detector outputs changes to "alarm", the overall output of the detector model is "alarm" as well.

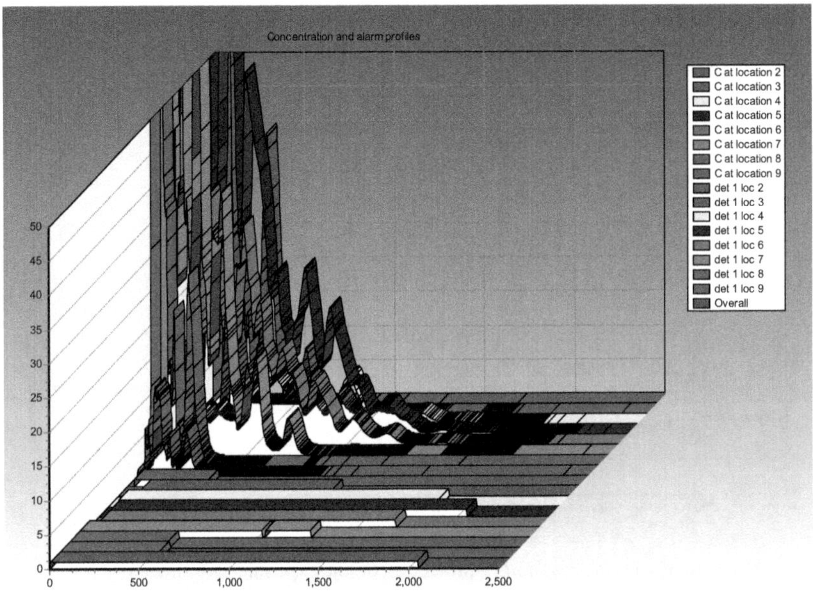

Figure 3. Concentration-time and alarm-time profiles of the individual detectors and the composed overall alarm profile.

Based on the concentration-time and alarm-time profiles, the skin and respiratory protection models calculate exposure profiles, using the characteristics of the provided protective equipment. The alarm is set almost instantaneously, see Figure 3. Therefore, the mask is worn immediately, only taking into account 15 seconds delay – the time it takes to put the mask on. Figure 4 shows the influence of wearing respiratory protection on the exposure to sarin. Clearly, the carbon filter provides sufficient protection in this case. However, in practice always leakage occurs to some extent through the mask. The figure shows two exposure profiles with different protection factors. As a consequence, the dosage that one inhales, or to which the eyes are exposed, varies with the actual protection.

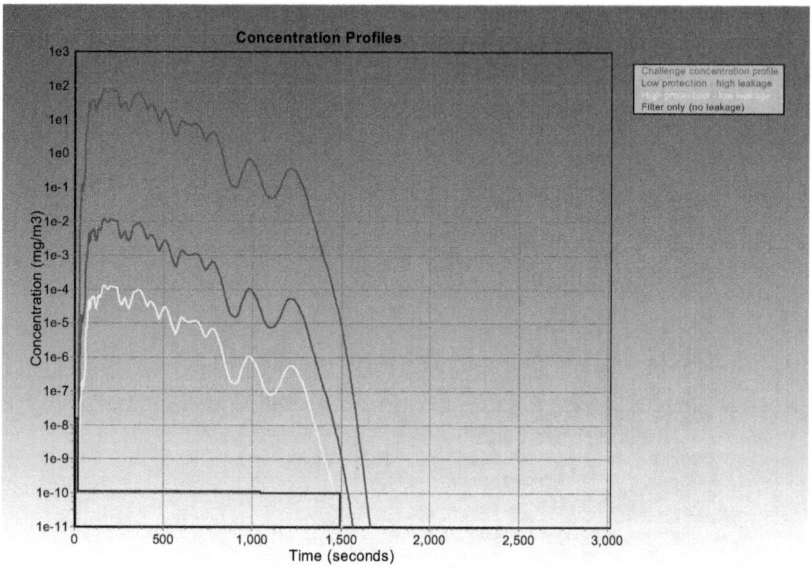

Figure 4. Influence of wearing respiratory protection on the exposure to sarin.

The toxic effects model translates the exposure profiles into casualty probabilities for the personnel, assuming a probabilistic dose-effect relationship. The casualty levels and spectra can be obtained for various type of health effects, e.g. eye effects, inhalation, percutane, subdivided in two levels (incapacitating and lethal), and various protection levels, e.g. no protection, suit only, mask only, mask and suit, and collective protection. Table 1 gives a typical result for one scenario. In case no protection is used, 63% of the population dies due to inhalation of sarin and 25% dies due to percutaneous exposure. Clearly, when both mask and suit are worn, the casualty levels are dropping drastically.

Table 1. Affected percentage of the population for various effects and protection levels.

Toxicity		No protection	Mask only	Suit only	Mask and Suit
Eye effects	Incapacitating	81	21	81	21
Inhalation	Lethal	63	4	63	4
	Incapacitating	76	11	76	11
Percutaneous	Lethal	25	25	2	2
	Incapacitating	45	45	13	13

All input parameters, scenario definitions and results are stored in a database for easy access and retrieval. Analysis of individual scenario results and statistical analysis over all scenarios (or any subset) is possible. Typical individual scenario results are deposition, dosage and casualty level on the attacked target. Typical statistical analysis results are dosage and deposition threat spectra, and casualty spectra. Figure 5 presents a typical example of a dosage spectrum. It shows the frequencies of occurrence that a certain dosage level is exceeded in 5% of the target area for three agents.

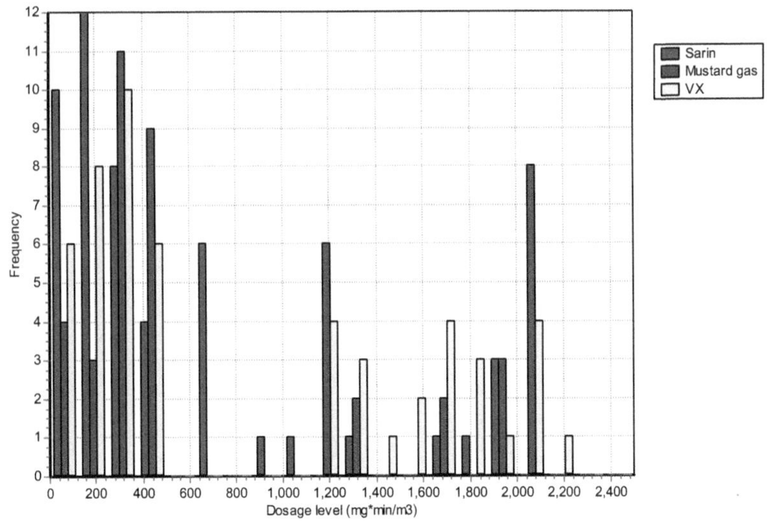

Figure 5. Frequencies of dosages that are exceeded in 5% of the target area.

Thus, the Chemical Incident Simulation model largely eliminates the subjectivity involved in scenario studies, and procurement of protective and detector equipment.

4. CONCLUSION

The example given above shows that CIS can simulate the effect of the complete passive defence chain in a consistent way. The strength of CIS is that it can simulate this effect for a huge number of different situations and thus is able to establish passive defence requirements in a systematic way. The proof of principle for simulating the complete protection chain has been given. Extensive work has to be done to refine this approach, so that CIS can be an effective tool to set requirements for real life situations.

This paper deals with the current status of this potentially very powerful tool and shows characteristics and capabilities, and some typical results. In the near future, while the CIS module will mature, it is foreseen that a so-called CIS user group will be initiated, which NATO countries can join.

Chapter 7

OPCW CONCEPT OF ASSISTANCE UNDER ARTICLE X

Clarence Lee Brown
Organisation for the Prohibition of Chemical Weapons (OPCW)

Summary: In conclusion the delivery of Assistance under Article X of the Chemical Weapons Convention is facilitated by the Organisation for the Prohibition of Chemical Weapons and the assistance received through the contributions of member states and the utilisation of experts within and external to the OPCW. It also involves the coordination and delivery of specialised services from national agencies and other international organisations involved in providing emergency humanitarian assistance. The OPCW will continue its work on the cooperative efforts with many member states to maintain the effort to development, implement and train for an effective delivery of assistance in accordance with the provisions of the Chemical Weapons Convention.

Keywords: ACAT; article X; antidotes; BAMS; chemical weaponce convention; conceps of assistance; detectors; mobile laboratory; OPCW, OSOCC;

1. DISCUSSION

The Chemical Weapons Convention (CWC) opened for signature 13th January 1993. Entry into Force was after the ratification of 65 signatory states on 29th April 1997. To date there are 167 ratified States Parties and 20 Signatory States which are yet to ratify. There are 16 Non signatory States which have yet to accede to the convention.

<u>Mission</u>

To implement the provisions of the CWC in order to eliminated an entire category of weapons of mass destruction, to Co-operate and used chemistry for peaceful purposes.

<u>Structure of the OPCW</u>

Conference of States Parties (All member states), yearly conference or as required, the Executive Council of 41 member states which meets regularly throughout the year and the Technical Secretariat of 506 staff members from 66 nationalities. The operating 2005 Budget is approx 76.5 million Euros.

C. Dishovsky et al. (eds.), Medical Treatment of Intoxications and Decontamination of Chemical Agents in the Area of Terrorist Attack, 71–79.

The Chemical Weapons Convention

The Chemical Weapons Convention (CWC) is an international treaty that bans the use of chemical weapons and aims to eliminate chemical weapons, everywhere in the world, forever.

The Convention provides the basis for the OPCW to monitor the destruction of existing stocks of chemical weapons and the facilities used to produce chemical weapons, as well as by checking many industrial sites to ensure that new chemical weapons are not produced. The OPCW also promotes international cooperation and the exchange of scientific and technical information, so that people and governments can benefit from the peaceful uses of chemistry.

By joining the Convention on the Prohibition of the Development, Production, Stockpiling and Use of Chemical Weapons and on Their Destruction (CWC), States Parties renounce any chemical weapons option. In return, States Parties may expect to receive the support of the international community through the Organisation for the Prohibition of Chemical Weapons (OPCW) if their security is threatened by the use or threat of use of chemical weapons. The provisions of this positive security guarantee will be an indispensable part of the Convention for as long as chemical weapons exist, for as long as States remain outside the Convention, and for as long as the threat of the use of these weapons by terrorist groups continues to cast a shadow over our lives.

In this context, the Convention requires that States Parties cooperate with the OPCW in facilitating the coordination and delivery of assistance and protection to minimise the consequences of a chemical weapons attack, in countering the threat of use of chemical weapons, and in eliminating the threats posed by activities prohibited under Article I of the Convention. Article X stipulates that each State Party has the right to request and to receive assistance and protection against the use or threat of use of chemical weapons.

This system is described in a concept and in annexes to the concept, and is the culmination of a review of all possible approaches to mobilisation. The concept and annexes propose a modular approach, with a limited stockpile at headquarters, and sufficient stockpiles within States Parties in various regions. These stockpiles would be ready to be mobilised within hours by one direct airlift. A limited stockpile could be deployed within hours after the receipt of a request, along with the OPCW's assistance coordination and assessment team (ACAT). The team would assess the needs of the requesting State Party, and on that basis, the assets stockpiled within States Parties would be dispatched. The Secretariat would ensure that the stockpiles met the necessary requirements, and were available when needed.

A reception centre and an on-site operation coordination centre (OSOCC) would be set up to deliver the means of assistance. Its backbone would be the ACAT team, reinforced by qualified experts and members of a protection network provided free of charge to the OPCW by States Parties. All of this would be facilitated through coordination with relevant international organisations, including the UN Office for Coordination of Humanitarian Affairs (OCHA).

The Organisation carries out a number of functions with respect to Article X.

General and routine functions:

To provide States Parties with information about a variety of protective measures against chemical weapons, as well as to receive such information as may be provided by States Parties;
To provide expert advice, and to assist a State Party, when requested, in identifying how its programmes for the development and improvement of a protective capacity against chemical weapons can be implemented[3];
To establish a voluntary fund for assistance, and to supervise and administer it[4];
To conclude agreements with States Parties in connection with Article X[5];
To coordinate the establishment and maintenance of permanent stockpiles of emergency and humanitarian assistance by States Parties, in accordance with subparagraphs 7(b) and 7(c) of Article X;[6]
To inspect the items for serviceability[7];
To compile declarations made by States Parties and register, when requested, bilateral agreements concluded between States Parties, or between a State Party and the Organisation for the implementation of Article X[8]; and
To assist States Parties, when requested, in developing emergency plans and legislation; in developing training programmes, procedures, and standards; and in developing an appropriate dispatching and receiving assistance mechanism.

Functions in times of emergency or crisis:

---

3 Article X, paragraph 5.

4 The Conference shall "establish at its first session the voluntary fund for assistance in accordance with Article X" (Article VIII, subparagraph 21(j)); The Executive Council shall "conclude agreements with States Parties on behalf of the Organisation in connection with Article X and supervise the voluntary fund referred to in Article X" (Article VIII, subparagraph 34 (b)); and the Technical Secretariat shall "administer the voluntary fund referred to in Article X" (Article VIII, subparagraph 39(c)).

5 Article VIII, paragraph 34(b).

6 The Technical Secretariat shall "...coordinate the establishment and maintenance of permanent stockpiles of emergency and humanitarian assistance by States Parties in accordance with Article X, paragraphs 7(b) and (c)."

7 Ibid.

8 Article VIII, subparagraph 39(c).

In the event of (a) the use of chemical weapons or riot control agents as a method of warfare, and/or (b) the threat of the use of chemical weapons, and/or (c) the threat of actions or activities prohibited for States Parties by Article I:

To make emergency, supplementary, and humanitarian assistance available to the requesting State Party, in accordance with the provisions of the Convention[9];
To coordinate the dispatch and delivery of means of protection, and to advise on the use of any of these protective measures[10];
To take emergency measures of assistance, using the resources the Conference of States Parties has placed at the disposal of the Director-General[11];
To transmit requests for assistance and relevant information;
To liaise with relevant international organisations for the purpose of communicating about reports on investigations of the alleged use of chemical weapons (IAUs); to coordinate the provision of assistance; and to obtain and exchange relevant information and data; and to conduct an IAU to establish the relevant facts related to the request for assistance[12].

Scenarios

With 167 States being States Parties to the Convention it is difficult to assume that a full-scale use of chemical weapons in an armed conflict between States should be used as a benchmark for assessing how adequate Article X offers actually are. If the Organisation is fully equipped for effective implementation of the provisions of Article X of the Convention, this will greatly contribute to overall compliance with the Convention, and immensely reduce the possibility of use of these weapons by States. Any potential aggressor using these weapons needs to have much larger capacity, and consequently much more visible programmes. It is impossible to completely rule out the military use of chemical weapons as a planning assumption. This leaves us with the possible scenario of CW use, and the threat of use, in a limited interstate military conflict, as well as the possibility of CW use in a terrorist attack. A State Party exposed to a chemical weapons terrorist attack has the right, under Article X, to request assistance to mitigate the effects of use. While the possibility of the use of CW between States is diminishing as membership in the OPCW increases, the threat of their use by terrorist groups has increased.

Preparation for "worst case scenario" needs

9 Article X, paragraphs 9-10.
10 Article X, paragraph 1.
11 Article X, paragraph 11.
12 Article X, paragraph 9.

Another method to determine the quantity of resources which are needed is to identify the requirements for a "worst case scenario". In this case, an attempt would be made to estimate the maximum impact which a chemical weapons attack would have. Such a scenario would involve a full-scale attack on a large city with an underdeveloped health system and infrastructure, and no preparedness or protective systems. The results could involve mass casualties. No organisation in the world, including the OPCW, would be able immediately to cope with such an event. The amount of resources to be stored, maintained, transported, and finally to be distributed, would be prohibitive. Additionally, although anyone can use general humanitarian assistance items such as food, tents, and blankets, the effective use of chemical protective items requires both instruction and training. The logistical requirements of such a worst case scenario would be overwhelming. Although such a scenario cannot be excluded per se, the probability of such an event is very low. Stockpiling items for such a scenario would be extremely costly, and would probably be less effective than the provision of a fully-developed protection programme to all Member States.

A modular approach to the mobilisation of resources

The modular approach is based upon the current capacity of the OPCW in terms of offers of assistance by States Parties (in the form of material and personnel), and the capacity of the Secretariat to store, maintain, transport, and distribute the items and resources most likely to be needed should a CW-related emergency arise. The components of this approach would involve a central stockpile of resources and equipment in the Organisation's headquarters, available for immediate dispatch to a requesting State Party, and in addition, stockpiles of resources which have been offered by States Parties and organised in the form of modules. The modules would be categorised into basic assistance modules (BAMs) for emergency assistance, and specialised assistance modules (SAMs) for supplementary and humanitarian assistance. The Secretariat would store BAMs and a very limited stockpile of humanitarian assistance, while both BAMs and SAMs would be stockpiled by the States Parties. The basic module could consist of pallet-packed items, ready to be airlifted within 12 hours after the request for assistance has been received. The BAMs in the headquarters' stockpile will be identical to the BAMs stored by States Parties. This will facilitate identification, packing, dispatch, and the delivery of assistance.

Headquarters resources

The concept of limited, centrally-located resources has been assessed, applying the assumption that a Secretariat team and the assistance coordination and assessment team (ACAT), are required to be in the requesting State Party within hours after a request for assistance has been received. This team will consist of a limited number of OPCW staff members, who will be carrying the equipment required for them to perform their own activities; in addition, they will be carrying a limited amount of equipment for the requesting State Party. It has been recommended that the Member States make provisions for a minimum internal capability within the Secretariat for the assistance needs of 1000 people (this is based on the standard payload of a C-130 Hercules transport plane). This limited central stockpile would be immediately dispatched, together with the ACAT.

Heavy equipment, units, and teams

In addition to the modular storage equipment offered by States Parties, there are heavy equipment items such as decontamination systems, mobile laboratories, field hospitals, medical equipment, and facilities for the temporary accommodation of persons. Protection equipment for the team will also be put into modules to facilitate storage, handling, and transport.

Medical Aspects of Assistance

In terms of medical assistance, a clear distinction is required between the role of the assisted State Party, the OPCW as a whole (inclusive of the collected medical abilities offered by Member States), and the Secretariat itself. Ideally, the requesting State Party will take responsibility for the actual delivery of collected services to the victims of a CW incident. The extent, to which the requesting State Party is able to offer medical services and take responsibility for the actual use of the medical assistance provided through the OPCW, will vary extensively from State to State. Where medical abilities and infrastructure are deficient, provision of medical services to victims will be required, using the resources which Member States have made available in their offers of assistance. In this case, it is important to identify responsibility for on-site operational command and control of the collected medical offers that may be available. This situation is analogous to the provision of comprehensive medical and health services in a disaster situation – a highly specialised function which should be the responsibility of specifically trained and experienced individuals. The Secretariat is medically staffed for the provision of an occupational health and safety service which supports OPCW inspections and headquarters activities, but it is not staffed or equipped for the provision of disaster medicine, or of a fully independent medical service. Even within this function, Secretariat medical staff are reliant on support from host States for the provision of comprehensive medical support. Where support is deficient, reliance on the rapid evacuation to regional centres of medical expertise is essential. The Secretariat's medical role in assistance activities is necessarily restricted to evaluating medical needs, coordinating the delivery of Member State medical resources to the site, ensuring that appropriate command and control of assisting medical units is provided (e.g. through the provision of a qualified expert with the relevant background in disaster medicine), and providing/coordinating advice/training to Member State medical personnel. Responsibility for on-site command and control of medical services should rest primarily with the requesting State Party, and only if this is not possible, will this be taken over from within the OPCW's assistance efforts.

Assistance coordination and assessment team (ACAT)

The main tasks of the ACAT are assessing the situation, advising and liaising with the RSP, instructing RSP personnel in the use of the items in the BAM, coordinating the receipt and delivery of assistance, and setting up the on-site operation coordination centre (OSOCC). The OSOCC may be set up by relevant international organisations such as UN-OCHA or a State Party (through an existing bilateral agreement, or an arrangement which has already been signed). However, the OSOCC should be staffed by the ACAT, which will be reinforced with outside experts and specialists. The ACAT will carry the equipment necessary for carrying out their activities.

The ACAT should consist of:

- The team leader;
- Assistance specialist(s) (who will also liaise with the SP);
- An NBC defence specialist;
- A safety officer;
- Medical doctor(s);
- Logistics specialist(s); and
- A communications specialist

The role and functions of the assistance response system

To establish a reception centre and an OSOCC in the requesting State Party by the ACAT, which will later be reinforced by experts and specialists from States Parties? The role of the OSOCC and its interaction with other relevant international organisations, assistance-offering States Parties, and the host country will be primarily in co-ordinating resources. This may be done, preferably through the conclusion of bilateral agreements and arrangements with these organisations and States Parties, and the requesting State Party. Although the OPCW has the obligation to provide assistance, the government of the requesting State Party and the local authorities are the ultimate decision makers, and are responsible for the receipt and distribution of assistance. The requesting State Party should specify as much as possible the scope and the type of assistance needed, and should identify its available local resources, such as experts, equipment, and materials. The requesting State Party may also request technical assistance from other intergovernmental or international organisations; and To conduct an investigation of alleged use of chemical weapons or riot control agents as a method of warfare, and to establish the relevant facts related to the request for assistance.

Reception centre and on-site operation coordination centre (OSOCC)

The role of the reception centre and of the OSOCC is to receive and distribute assistance. The ACAT of the OPCW, with the help of its external experts and in cooperation with States Parties and other relevant international organisations, will set up the centre and the OSOCC. There are States Parties and organisations which are willing to set up the OSOCC for the OPCW. This operation, however, requires conducting field exercises and training with the offering States Parties and international organisations.

Relations with other international organisations and national agencies

The OPCW will cooperate with relevant international organisations for the delivery of assistance. Such a recommendation is based on a lack of resources within the OPCW to individually provide delivery of assistance, and the fact that it is not economical for the Organisation to seek a delivery capability. In addition, cooperation with international organisations and national agencies helps avoid a duplication of activities, particularly on the site of operations; furthermore, the Organisation can utilise the experience of other international organisations.

Areas of cooperation with other international organisations and national agencies may expand to systematic information monitoring in certain cases; these cases could provide advance warning of a situation which might eventually result in an official request for assistance (early warning). Daily monitoring of international news, and the regular use of specialised databases, are possible through such cooperation. Furthermore, this approach enables the OPCW to have access to the experts and specialised expertise of other organisations.

An analysis of the Convention's requirements and available resources indicates that cooperation with other organisations is required in at least the following four areas:

- In coordinating the dispatch of assistance;
- In the delivery of assistance;
- In managing on-site activities; and
- In training.

Chapter 8

ENVIRONMENTALLY HAZARDOUS PROJECTS IN UKRAINE AS ATTRACTIVE TARGETS FOR TERRORIST ACTS

G. Shmatkov

Ecological Association of Ukrainian, Mining & Smelting Complex Enterprises, Dnepropetrovsk, Ukraine

Environmentally hazardous projects are those where the risk of accidents is very high, which can result in a major and sometimes even catastrophic chemical pollution of the environment. Frequently, these disasters take casualties among the plant personnel, as well as among the nearby settlements population, which were the cases with the Chernobyl Nuclear Power Plant disaster in Ukraine, or with the pesticide plant accident in Bhopal, India.

Let us have a look at the respective situation in Ukraine. Ukraine is the biggest European country bordering with the European Union (EU) countries. Its territory is bigger than that of any individual EU country. The area of Ukraine is 603,700 square kilometers, while the area of France is 543,965 km^2, and that of Germany is 356,733 km^2. Economy of Ukraine is based on mining and heavy industries, such as metallurgical, chemical, heat-and-power, and heavy engineering branches. These branches of industry are known to be the most hazardous for the environment and have a negative impact on all its components: atmosphere, water resources, the Earth's interior, soil, flora, and fauna.

These projects annually stockpile millions of tons of waste materials falling under Danger Grades I through IV, which are accumulated in special precipitation tanks, hoarders, tailing dumps, whose volumes often amount to hundreds of thousand and millions of cubic meters. The tailing dump levees at many ore mining and processing enterprises rise hundreds of meters high. If these levees happen to burst or collapse, there is a threat of powerful mud slide torrents which can spread over the nearby settlements and get into land surface reservoirs, as most of

C. Dishovsky et al. (eds.), Medical Treatment of Intoxications and Decontamination of Chemical Agents in the Area of Terrorist Attack, 81–83.

those giant hoarders are located near water bodies, in ravine beds and small rivulets flowing into the Dnepr.

The most dangerous in their long-term effects are the tailing dumps containing radio-active waste, which are also located in the vicinity of the Dnepr bed or near smaller rivers in the Dnepr catchment basin. Many tailing dump levees and toxic & radio-active waste hoarder dams are in extremely poor condition. They can be easily destroyed by natural factors (minor earthquakes, floods, rise of subterranean waters) or by man-made acts (like a terrorist act with the use of explosives), with all ensuing consequences.

Another class of environmentally hazardous projects comprises finished stores of toxic, explosive, and flammable substances. In Ukraine, it comprises depositaries of benzene, petroleum, and oil products, ammonia, inorganic acids, rubber resin, pesticides, mineral fertilizers, and many other materials. The quantities of substances kept in those stores amount to tens and hundreds of tons. Many of them are situated near beds of potable water bodies. Besides being environmentally hazardous, they present attractive targets for terrorist acts, since in case of their damage the chemical contamination of the environment may spread over vast territories and water areas creating an extremely dangerous situation for the population.

There are over 400 potentially hazardous industrial projects in Ukraine, 27 of which have the highest risk of accidents disasters with catastrophic consequences. During the last three years there have been registered 74 emergency situations of natural character, while there have been 123 records of man-caused emergencies, which is almost twice as many. This is a vivid illustration of the extremely poor technical condition of potentially hazardous installations.

The third potentially most hazardous kind of projects is trunk pipelines. There are trunk pipelines in Ukraine that are tens, hundreds, and even thousands of kilometers long, through which hundreds of thousand and million cubic meters of oil, gas condensate, gas, ammonia, toxic chemical waste, mineral ore-dressing waste, including radio-active one, are pumped over long distances.

The main-laying systems are most diverse, - underground, overground, underwater, and water-surface ones. Many of them pass near towns and villages, cross the beds of small and big rivers, including the Dnepr, approach the Black Sea coast.

Damage of these pipelines may result in large-scale contamination of water bodies, subterranean water-bearing (including potable water) horizons, and fertile soils. Some of those substances are volatile (for example, ammonia) and can affect the population through the atmosphere. These very long-distance pipelines often passing through deserted areas may serve as a vulnerable target for terrorist acts, with catastrophic consequences in respect of the chemical contamination of the territory, water bodies, and population.

At present it has become exigent to work out a governmental program in Ukraine for securing environmentally hazardous installations that might be attractive for terrorist acts.

Chapter 9

DIOXINES: THREAT OF MISUSE IN POSSIBLE ACTS OF CHEMICAL TERRORISM

V. F. Tkach, N. G. Prodanchuk, N. V. Kokshareva, M. L. Zinovjeva
L. I. Medved' Institute of Eco-Hygiene and Toxicology, Kiev, Ukraine

Keywords: carcinogenicity; chemical terrorism; chlororganic production; dioxins; fibrosarcoma cutis; porphyria; TCDD; toxic action

Because of very high toxicity, ability of delayed action and high stability in the environment, polychlorinated dibenzodioxins (PCDD), dibenzofurans (PCDF) and biphenyls (PCB) are particularly hazardous for people's health if misused for committing acts of terrorism, or under the circumstances of violating the rules of toxic wastes' safe storage.

"Dioxins" is a generalized name of the large group of polychlorinated dibenzoparadioxins (PCDD) and polychlorinated dibenzofurans (PCDF).

Depending on quantity and location of chlorine atoms, quantity of isomers of these compounds totals 210, including 75 isomers of PCDD, with only tetrachlordibenzo-p-dioxins (TCDD) accounting for 22 isomers [1,2].

In the narrower sense, dioxin term is assumed for 2,3,7,8-tetrachlordibenzo-p-dioxin (2,3,7,8-TCDD) only. 2,3,7,8-TCDD is odorless and colorless non-hygroscopic powder, featuring high adhesion ability and static behavior.

Empirical formula is $C_{12}H_4Cl_4O_2$. As a rule, 2,3,7,8-TCDD is combined with other isomers in chemical substances, bio-media and objects of the environment.

Dioxin density at 25°C is 1,827 g/cm^3; boiling point at 1 atmosphere is 421,2°C; melting point is 305°C; and decomposition temperature is 800°C. Dioxin is virtually insoluble in water (8-200ng/l); it is dissolved in organic solvents: o-dichlorobenzene, chlorobenzene, tetrachloroethylene, chloroform, hexane etc. Solubility in fats makes 44mg/kg. Dioxin is thermally stable; it is not oxidized by atmospheric oxygen, not hydrolyzed in water, stable to action of strong acids and bases. Dioxin quickly decomposes in the solutions of organic solvents

C. Dishovsky et al. (eds.), Medical Treatment of Intoxications and Decontamination of Chemical Agents in the Area of Terrorist Attack, 85–89.

under UV-radiation action, forming less chlorinated and less toxic derivatives.

Dioxin half-life period under photolysis is: on glass - 6 days, in methanol - 6 hours, in hexane - 1 hour.[3, 8].

PCDDs are unavoidable companions of chlororganic production processes; as by-products, they are formed in micro-quantities in the course of any chlororganic synthesis, and come in the environment mainly with production wastes. Disposal of these toxic wastes requires specially equipped storages. In real conditions, thousands tons of sludge and mud containing PCDD and PCDF, are lumped on the territories of chlororganic synthesis enterprises in sludge reservoirs and drying beds where they are stored open for decades, exposed to various environmental factors and presenting a threatening source of secondary pollution of subsoil waters and atmosphere [4].

Incineration of domestic waste is also a contributor to environmental pollution. One more source of dioxins is pulp-and-paper industry. Comparatively new ways of forming polychlorinated dibenzo-p-dioxins and furans are high-temperature processes like copper melting in electric arc furnaces, and production of magnesium, nickel and, possibly, other metals of their chlorides.

Heating of houses with wood, charcoal or mazut also results in formation of trace amounts of dioxins. Hazardous source of PCDD and PCDF constitutes the combustion of engine oil and gasoline with chlorine- or bromine-organic additives. Over the long period of time, pesticides containing dioxins (2,4,5-T, copper trichlorophenolate etc.) [4] were used in agriculture.

Dioxins are extremely toxic compounds. LD50 for the most toxic dioxin (2,3,7,8-TCDD) at intragastric injection was 0,6-2,0 mcg/kg for guinea pigs, 3000-5000 mcg/kg for various species of hamsters, about 70 mcg/kg for monkeys and 25-45 mcg/kg for rats.

Maximum non-effective doses for rats were established too. On general toxic action, this dose is equivalent to 0,00001 mg/kg a day during 13 weeks. As regards carcinogenic effect and impact on reproductive function, it is 0,000001 mg/kg. For monkeys, maximum non-effective dose as to teratogenic effect makes up 0,00005 mg/kg.

Apart from general toxic action, dioxins, in particular 2,3,7,8-TCDD, cause thymus involution and immunodeficiency, however, less pronounced than in the event of HIV infection. Experiments with mice have shown that cell-mediated immunity was suppressed even at doses of 0,04 mcg/kg.

At doses of 0,01-0,05 mcg/kg, dioxin has mutagenic effect on lymph cells in mice. Embryo- and gonado-toxicity, as well as teratogenicity, are also observed.

Dioxin carcinogenicity for rodents and primates was proved by experiments. In rats, tumors of adipose tissues are mainly observed; in mice, skin application of dioxin causes fibrosarcoma cutis at the place of exposure [3].

Hazard of dioxin-like compounds is growing because they can be accumulated in living organisms when coming from the environment.

The main depot for dioxin accumulation in mammals is the liver and adipose tissue. When dioxin is fed to rodents with fodder, about 10% of the dose comes in the liver. Should this toxic agent be applied on rat skin, 13% of the dose is absorbed in blood already in 5 hours, and 10% remains on skin surface.

In mice and hamsters, dioxin is excreted with urine, in the rest of animals – with faeces.

Compounds of this series are actively accumulated by the inhabitants of water ecosystems (both plants and animals). It was shown that fish in rivers, polluted with dioxin-containing industrial wastewaters, could accumulate up to 0,1-0,2 mcg/kg. As much as 80% of dioxin is laid in fish skeleton and visceral fat; about 10% comes in liver, and 2% - in muscles.

Tubers and edible roots are featuring the highest dioxin-accumulation ability. For example, carrots can extract dioxin from soil and accumulate it up to concentrations being equal to, or somewhat exceeding, its content in soil.

Dioxin exceeds by toxicity all the known synthetic poisons; it is 2000 times more toxic than strychnine. Rated lethal dose of dioxin for humans is 70 mcg/kg, and minimum effective dose totals 1,0 mcg/kg of body weight. Poison is introduced into human body through lungs, digestive tract, and undamaged skin, causing diseases of prolonged and flaccid course developing in 2 - 4 weeks and more.

Dioxin half-life in human body can vary from 1 to 8 years.

Background dioxin content in adipose tissue of humans is 4-6 pg I-TEQ/r (fat), and that value in blood is 0,9 - 480 pg I-TEQ/r (lipids). These indices depend on proximity of pollution sources to places of people's residence and degree of their contact with objects polluted with dioxins.

Acute dioxin poisoning of humans is characterized by latent period lasting 1-4 weeks after poison is getting into the human body [3].

Dioxin is a poison of polytropic action, affecting actually all the organs and systems of the body. The most typical manifestations of dioxin poisoning are skin affections, such as chloracne.

Chloracne is highly specific sign, occurring under contact with chlorinated cyclic organic compounds only. Period of affection lasts from several months till several years, depending on degree of poisoning.

In humans, chloracne is provoked also by chlorinated benzofurans and biphenyls, chloronaphthalenes, tetrachlorazobenzene and tetrachlorazoxybenzene.

Chloracne is acne-like eruptions located on upper eyelids and under eyes, on skin of cheekbone part of cheeks, on the other side of ears, on the nose. Rather seldom, eruptions spread to skin of the axillary creases, inguinal region, chest, back, and hips. Appearance of chloracne is preceded by edema, erythema of skin. Eruptions are accompanied by itching. Pyoderma and abscesses can also develop. Further, colloidal scars and hyperkeratosis are formed. Liver affection is expressed to a variable degree: from short-term swelling of liver without any pronounced functional disorders to severe abnormalities in liver functions and structure, down to necrosis. Content of alanine and asparagine transferase in blood serum increases; fat exchange and carbohydrate metabolism deteriorate. Pathology on the part of nervous system is often observed – from subclinical neuritis revealed with the help of electromyography to polyneuropathy and polyneuritis. Impairment of hearing, olfactory and gustatory sensitivity can likely happen. Mental abnormalities, as well as asthenic and depression syndromes, are also not uncommon.

As regards respiratory apparatus, intoxication is manifested in the development of catarrh of upper air passages and bronchitis with dyspnea, but more often damage is limited to impaired vital capacity of lungs. Affection of gastrointestinal tract is manifested in gastritis and chronic colitis.

In some cases, muscle pain, edemas in extremities, bursitis in the area of large joints are observed. Severe cases of acute poisoning may lead to anemia, leukopenia and even pancytopenia.

Under chronic intoxication, late skin porphyria is added to the above diseases. This syndrome is characterized by increased excretion of carboxylated porphyrins, coverlet pigmentation, photosensitivity, intes-tinal and nervous disorders. Toxic carboxylated porphyrins and ions of iron, not included into porphyrin structure, are accumulated in the cells of various organs.

Late porphyria, as a rule, is accompanied by swelling of the liver, its abnormal functioning, change in color of urine to dark-orange.

TCDD is characterized by various toxic manifestations, with pronounced interspecific and inter-tissue differences, both in terms of quantity and quality.

TCDD toxicity, obviously, depends on the fact that four lateral positions in the molecule are occupied with chlorine. Toxicity is reduced with lateral replacement and increase in total replacement with chlorine. [3].

Comparative analysis of results of assessing state of health of people exposed to “Orange Agent” action during chemical war in Vietnam allowed detecting long-term effects of dioxin. Data of clinical examination of 332 Vietnamese veterans prove the existence of new pathology, characterized by abnormalities in cardio-vascular, respiratory, digestive, nervous, urinary, musculo-skeletal and reproductive systems, as well as reduction of life span [5].

Mechanism of 2,3,7,8-TCDD was not established so far; means of specific therapy as to this compound poisoning are not available. Experiments with animals have shown that activated carbon, zeolite (subject to introduction of sorbents immediately after poison), unithiol, Liv-52, carsil, festal, guaranteed survival of 20-50% laboratory rats [6].

There is a great number of dioxin-hazardous chemical enterprises, non-ferrous metallurgy plants, and works engaged in thermal treatment of industrial and domestic wastes, on the territory of Ukraine. Taking this fact into consideration, a threat of polluting the objects of the environment with dioxins exists.

The Institute of Eco-Hygiene and Toxicology (Kiev) is using the device Polaris Q (Termo-Finnigan) for screening and analysis of dioxins in ms/ms mode in soil, ash, water and fish species. 17 isomers (furans and dioxins) are determined with high-resolution mass-spectrometer MAT-95XP according to the method EPA-1613.

Investigations have shown that plants on thermal treatment of solid domestic waste represent the source of polychlorinated dibenzodioxins and biphenyls. In this connection, it is necessary to conduct further investigations of their emission and determination of all the range of PCDD, PCDF and PCB [7].

REFERENCES

1. A. F. Troyanskaya, D.P.Moseeva, N.A.Rubtsova. Ecological Consequences of Using Sodium Pentachlorphenolate at Woodworking Enterprises in Arkhangel'sk Region. – In the book:
Dioxins – XXI Century Supertoxicants.- Regions of Russia.- News bulletin No.3.- Moscow, 1998, p.1-9.
2. R. I. Pervunina, D. P. Samsonov, V. P. Kiryuhin. Studying of Environmental Pollution with Dioxin and Related Compounds in Cities and Regions of Russia (1988-1996). - In the book: Dioxins – XXI Century Supertoxicants.- Regions of Russia.- News bulletin No.3.- Moscow, 1998, p.64-81.
3. Hygienic Criteria of the Environmental Conditions 88. Polychlorinated Dibenzo-para-dioxins and dibenzofurans. - World Health Organization (WHO).- Geneva.-1993.- 381 p.
4. I. M. Muslimova, F. F. Hizbullin.- Problem of Accumulation of Polychlorinated Dibenzo-p-dioxins and Dibenzofurans in Sludge Reservoirs of the Enterprises of Chlororganic Synthesis. Ecological Chemistry.- 2001.- Vol.10.- No. 04, p.269-274.
5. A. V. Epifantsev – Long-term Effects of Dioxin-containing Eco-toxicants: Dioxin Disease. In the book: Second Congress of Russian Toxicologists. - 2003.- Moscow, p. 99-101.
6. M. G. Lavrov. Effect of Pharmacological Preparations on 2,3,7,8- TCDD Intoxication. Second Congress of Russian Toxicologists. - 2003.- Moscow.- p.357-359.
7. M. G. Prodanchuk, V. I. Medvedev, R. Yu. Sova. – Evaluation of Threat of Environmental Pollution with Dioxins and Biphenyls under Thermal Treatment of Solid Domestic Waste. In the book: Cleanness of the Environment in Our City.- Theses of Reports.- Ukraine, Lviv.- 1999.-p.16-18
8. Yu.A.Zolotov. Dioxin Detection and Determination. Journal of Analytical Chemistry. - Nauka.- Publishers: MAIK "Nauka / Interperiodika".- Vol. 57.- No. 2.-2002.- p.117.

Part 2

DIAGNOSIS OF EXPOSURE TO CHEMICAL AGENT AND MEDICAL TREATMENT OF CHEMICAL AGENT INTOXICATION

Chapter 10

EPIDEMIOLOGY OF CHEMICAL CRISIS, PUBLIC HEALTH IMPACT, SPECIFIC MEDICAL COUNTERMEASURES AND EDUCATION

Florin Paul[1], Liliana Paul[2]
[1]*Romanian Armed Forces, Ministry of National Defence, Bucharest, Romania*
[2]*Vocational High School "Sf. Pantelimon", Bucharest*

Abstract: Chemical attack and/or terrorism is the intentional use of chemicals to scare, injure, or kill people. Although rare, chemical agents have been used to disrupt daily activities, such as the 1995 release of nerve gas in a Japanese subway. Sometimes hazardous chemicals are also released accidentally; it may be very difficult initially to determine intent. Depending on the nature of the chemical(s) and the manner in which they were released, chemicals can contaminate the air we breathe, the water we drink, the food supply, or surfaces that people contact. Methods to spread chemicals may be as simple as opening a container, using a common garden sprayer, or as tricky as exploding a bomb with chemicals inside.

Harmful chemical exposures are usually characterized by rapid onset of medical symptoms (minutes to hours) and easily observed signs like unusual coloured residue, odours, dead or dying plants, insects, and animals.

Exposures to hazardous chemicals may cause a wide range of possible health effects depending on the nature of the chemical(s) used, and a number of other factors.

The pressure on the public health services in such events is extremely larger and without prior preparedness the situation could not be controlled.

Planning the resources, stockpiling the medicine, training the responders education of population and especially young generation , and preparing the community are essential for an appropriate response in chemical crisis.

Keywords: Hazardous chemical substances; Chemical Accidents; Chemical Terrorism; Risk Assessment in Chemical Threats; Public Health; Education.

C. Dishovsky et al. (eds.), Medical Treatment of Intoxications and Decontamination of Chemical Agents in the Area of Terrorist Attack, 93–99.

A chemical crisis could occur in many circumstances, ranging from industrial incidents or accidents to criminal use of chemical substances against humans, animals and the environment.

A very special and sensitive issue is the production and usage of the warfare chemical agents. The Chemical Weapons Convention (CWC) which now has 168 member states aims to rid the world of all chemical weapons by 2007. Now three years before the Convention's end date, the promise is seen optimistic.

The extremely large variety of chemicals and the wide range of their usage in day by day life made it practically impossible to keep an effective control of their production, stockpiling, transportation, selling and usage.

Large amounts of hazardous toxic industrial chemicals and substances are produced, stored, manipulated and transported all over the world, day and night.

In the mean time many of them are concentrated at certain sites, such as industrial zones, large stockpile barrels and reservoirs, railway stations etc.

For example hundreds or thousands stores with ammonia may be found every where in the world, as it is used in food processing facilities such as breweries, dairy works or slaughter works.

There exist real possibilities for the misuse of these chemicals like wrong stockpiling, leaking or accidents.

Industrial chemical substances are in many cases dangerous due to their inflammability, explosiveness or poisonous effects. They may be , oxidizing, irritating, carcinogenetic and mutagenic or dangerous for the environment; many other of them may spontaneously react with water, or generate toxic gases with water.

The main impact of the chemical accidents and chemical terrorism may differ by the chemical involved in event. However, three basic events could be counted:

- fire
- explosion
- dangerous substances leaking and dispersal.

These may have four destructive impacts that have negative influence on humans, animals, environment and property:

- heat radiation
- pressure wave
- debris
- toxicity

From all of these the most dangerous for the humans are toxic agents and substances.

The toxic substances leakage could cause deadly or harmful effects on humans, fauna and flora. Their dispersal depends on many parameters:

- type of agent
- amount
- physical state of the chemical
- mode of leaking or escape
- meteorological conditions

The high risk sources for chemical accidents are:

- industrial plants, usually located near inhabitated areas
- rail tankers and railway station
- road tankers and highways and large parking areas

All of these may be targets for the terrorist attack, with high impact in the population.

The local authorities must inform the population in these risky areas about the dangerous risk sources and about their potential impacts, especially in the vicinity of such sources and facilities.

The population must be informed about the methods of warning, and the way of behaviour in case of accident.

It is, therefore, ideal and necessary for each person living in risky areas to have particular knowledge about the chemical accidents and the first aid, the first action needed if the accident occurs, the evacuation way etc.

Unfortunately the time we are living now brought the terrifying shadow of terrorist attacks, and Nairobi, Tokyo and New York already experienced them.

The terrorist attack is very difficult to be predicted and it could be performed everywhere, anytime. A wide range of weapons and methods could be used by the terrorists to achieve their goals. Explosives, guns, chemical and biological agents may be used. Water and food contamination, pollution of air controlled systems and many others could be listed, and the imagination has no limit, as shown by the 9/11 attack.

Thinking about the chemical terrorism, it is necessary for the terrorists to know how to prepare and perform the attack.

Toxic chemicals that are easy to be found in the industrial areas are: chlorine, ammonia, hydrogen cyanide, formaldehyde, phosgene, hydrogen chloride. All of them are very harmful upon dispersal in the environment and would cause acute syndromes that require immediate qualified medical care.

Epidemiological profile of chemical crisis has specific aspects, pending on the nature of the crisis.

In the industrial incidents or accidents the basic facts taken in consideration are:

- chemical substances produced or processed;
- technical risk of the installation;
- distance from the inhabitant areas;
- meteorological conditions;
- transportation and route to and from the plant;
- evacuation capabilities both from the plant and city;

- population in the high risk area;
- estimation of casualties
- medical facilities
- etc.

In the terrorist attack many other items must be considered, as the follows:

- way of acquisition of the chemical:
  - robbery (stealing the road or train tanker);
  - illegal purchase;
  - clandestine fabrication
- initiation of the hazardous chemicals directly at source;
- amount and quality of used toxic chemical substances;
- knowledge and skills of using industrial toxic substances;
- selection of the appropriate site for the terrorist attack;
- selection of the appropriate time for the terrorist attack;
- availability on the market (legal or "black") of the technical equipment and basic components to prepare disposal devices or toxic substances
- financial resources required to prepare the attack
- ability of the terrorist to use the local characteristics: population density and habits, meteorological, transportation, power plant, communication etc.

Looking at the above list, the conclusion is clear: the response and the efficient action in case of accident need a very good previous plan and preparedness activity.

There are no doubts that this problem must be handled by the local authorities, with strong governmental support.

A preparedness plan must cover all the activities and measures needed to be taken to reduce as much as possible the effect of the attack and to save the live of the casualties.

A comprehensive plan will cover the following issues:

1. Protection of Objects and Facilities

Strong legal provisions and standard procedures in chemical industry and chemical handling must be in force.

Regime precautions compose an aggregate of the rules for securing the safety of objects or installations. Electrical security system, access and departure into facility of employees and vehicles, fire alarm signalization etc.

2. Public Health and Medical:

By using data on population density in the vicinity of the particular risk sources, it is relatively easy to estimate numbers of dead and wounded/affected citizens, and even livestock.

The mechanisms of each hazardous chemical's toxic effects differ and depend on many factors. The majority of substances take effects through inhalation, but also by ingestion. Both ways are very easy to be used by terrorist in their attempt.

The following issues may define the impact of the Chemical Accident/Attack on the public Health System and to estimate the amplitude of the event:

- Resources available vs. amplitude of the attack!
- Local & regional capabilities (the wide range of them – comms, transport., drugs)
- Manpower and qualified personnel
- Financial resources
- Realistic & adequate planning and training

There are no doubts that the overwhelming of medical resources occurs every time.

Due by that a realistic and appropriate estimation of the required resources must be based on the risk assessment in the high risk areas.

Specific medical countermeasures must be prepared and probably the main factors contributing to that are listed below:

- Nature of chemicals and their effects on humans;
- Mechanisms of action against health and environment
- Way of entrance : respiratory, ingestion, cutaneous
- Population density
- Hospitalization capabilities and technology
- Availability of the antidotes, general drugs and the way of administration – devices needed
- Qualified personnel
- Who replaces Medical Personnel injured in the incident?

All of these questions hide a very complex and expensive planning activity that can not be achieved in many developing countries. However a well tailored plan to the local capabilities must be created and put in place and followed by continue training programme.

3. Education and Training

Increased public knowledge is probably the best defense or response to the new chemical weapon threat.

The ideal case could be a public knowledge campaign to teach people what the threats are and how to identify the signs of a chemical weapon attack. For the time being, education can be on the managerial level. Large corporations, organizations, schools may designate people to become educated in what to do in the event of such an emergency. These people could in turn instruct the masses of people in an emergency. Likely targets of chemical attacks could also instruct their personnel and post signs regarding the response to such an emergency.

This increase in awareness would most likely contribute greatly to a reduction of chemical weapon casualties.

The education of the young people is essential, not only for themselves but also for the entire community. The acknowledgment will bring in the population a new level of understanding and in subsidiary a better approach of the problem.

It is recommended to develop a Scholar Curricula on Bio-Chemical Crisis.

This will consist in basic knowledge about the treats, chemical compounds and their effect on the health, biological pathogens, preventive measures. The main aim is to reduce the fear and panic and to instruct people how to behave in case of emergency.

Scholar Curricula must cover:

- Basic knowledge about chemicals, biological agents and harmful substances;
- Lessons about the first aid in case;
- How to cope with fear and panic

Nevertheless with out a continuous training any action is less effective.

CONCLUSION

Industrial accidents and terrorist using hazardous chemical substances and preparations can easily occur.

Fast and reliable modelling of the accident impact, not only a major accident but also of chemical terrorism, is important for early and correct protection of the affected community, from warning through protection to treatment.

The complexity increases when we look at a mobile risk source with hazardous chemical substances, where transport routes are not usually know.

Among the issue with mobile risk sources are , for example, transport of high volumes of hazardous substances through densely populated areas, parking the tanker vehicles at improper sites, passage of the tanker vehicles through the spots with the highest occurrence of car accidents etc.

The authorities must also allow suitable and full-scope information to the impacted population about the possible risk sources, their impact in case of accident, methods of warning and recommended models of behaviour under the occurrence of such extraordinary situation.

And the last but not the least, education of the young generation will bring a plus of confidence of the community facing with threats.

REFERENCES

1. Mika o., Vanecek M. Foxing the toxic terrorists. Defence International, winter 2004, 74 - 77.
2. C.D.C. The Public Health Response to Biological and Chemical Terrorism. Centers of Diseases Control, U.S., 2001.
3. Marcko S. The Threat of Chemical and Biological Attack, Emergency Net News,1996,08/27;
4. CDC. Biological and Chemical Terrorism: Strategic Plan for Preparedness and Response. MMWR, vol.49, 21 April 2000;
5. Godber A. Terrorist Use of Chemical Weapon. http://www.uberhip.com/people/ godber/index.html;
6. Kenyon Ian. Chemical Weapons in the Twentieth Century: Their Use and Their Control. The CBW Convention Bulletin, 48, June 2000;
7. Pearson G. S. Toxic Chemical: A Multinational Import Export System. The CBW Convention Bulletin, 34, Dec. 1996

Chapter 11

MODERN APPROACHES TO MEDICAL TREATMENT OF POISONING CAUSED BY NEUROPARALITIC ANTICHOLINESTERASE COMPOUNDS

Kokshareva N.V., Prodanchuk M.G., Tkach V.F., Zinovieva M.L.
Medved's Institute of Ecohygiene and Toxicology, Kiev, Ukraine

Abstract: As means for terrorism, chemical, biological, physical, technical and other means can be used. Their usage makes up world threat for the development of eco-toxicological catastrophes. Chemical substances hold an important position among the terrorist means. Particularly, it can be military chemical agents, which can fall into terrorists' hands by force of either taking or damaging the places of their storage as a result of armed attacks. Besides, the usage of chemical substances of other destination (such as industrial chemicals, pesticides, etc.) cannot be excluded because of their greater accessibility. The usage of modern knowledge, training, hardware and financial resources by a subject appeared for a terrorist does not exclude the synthesis of toxic substances by underground laboratories.

Among chemical substances that are potential substances for terrorist actions, a special place belongs to cholinesterase blockers, particularly to organophosphorous compounds, which are either in arsenal of many countries (sarin, soman, tabun, V-gases, etc.) or are used as plant protectants against diseases and pests. The usage of some of them even in small amounts can affect a large contingent of the population, cause death or invalidization, and do damage to the environment as well.

Mechanisms of toxic action of organophosphorus compounds as potential agents for chemical terrorism and the contemporary approaches to therapy of acute poisonings induced by blockers of cholinesterases will be presented in the report. It is known that, of all anticholinesterase compounds, organophosphorus and carbamate substances inhibit cholinesterase irreversibly, with the former being most danger in respect of both acute poisonings and delayed neurotoxic effect. Represented are cholinesterase reactivators of different generations, cholinolytics and other agents for the treatment of acute poisonings and prophylaxis neurotoxicity in poisonings with organophosphorus substances. Some perspective directions are substantiated for both the development of novel combined antidotes (reactivator-cholinolytic) and the search of efficient therapeutic means for acute poisonings with organophosphorus substances.

C. Dishovsky et al. (eds.), Medical Treatment of Intoxications and Decontamination of Chemical Agents in the Area of Terrorist Attack, 101–108.

Keywords: anticholinesterase compounds; ChE reactivators; delayed polineuropathy; medical treatment; poisonoing

1. DISCUSSION

At present time, mass terrorism acquires a progressively growing scale all over the world. As a result of acts of terrorism, a lot of people perish, and extensive damage is caused to the ecosystems. Scales of terrorism threat can be evidenced by the known cases in Matsumoto (1994), Tokyo (1995), Amman (1997), New York (2001), Moscow (1997, 2002, 2003), etc.

During the last decades, mass terrorism has developed into international terrorism. Known are transnational terrorist groupings, particularly “Aum Shinrikyo”, “al Qaida” and others, which are able to use any methods of mass terrorism to demolish people, animals, vegetation, stocks of materials and capital equipment.

Chemical substances hold an important position among the terrorist means. Particularly, it can be military chemical agents (MCA), which can fall into terrorists’ hands by force of either taking or damaging the places of their storage as a result of armed attacks. Besides, the usage of chemical substances of other destination (such as industrial chemicals, pesticides, etc.) cannot be excluded because of their greater accessibility. The usage of modern knowledge, does not exclude the synthesis of toxic substances by underground laboratories.

Among chemical substances that are potential substances for terrorist actions, a special place belongs to cholinesterase (ChE) blockers, particularly to organophosphorous compounds (OPC), which are either in arsenal of many countries (sarin, soman, tabun, V-gases, etc.) or are used as plant protectants against diseases and pests. The usage of some of them even in small amounts can affect a large contingent of the population, cause death or invalidization, and do damage to the environment as well.

Based on features of their interaction with ChE, anti-cholinesterase substances are classified as reversible and irreversible inhibitors. Referred to the reversible anti-cholinesterase substances are quaternary ammonium compounds and aminoformic acid esters; OPC are referred to the irreversible anti-cholinesterase substances.

The mechanism of interaction between such compounds with ChE and cholinoreceptors (ChR) has been studied in detail.

OPC manifest toxic effects as a result of certain structural similarity with natural ChE substrate, acetylcholine (ACh), by both stereochemical features and reactivity.

Dose-effect dependence is shared by many OPC in both the acute and chronic tests. The higher dose of an anti-cholinesterase substance, the higher degree of both acetyl cholinesterase (AChE) inhibition in neural tissue and intoxication evidence.

In most cases, the appearance of first cholinergic signs and symptoms is observed at the time when AChE activity in blood is reduced down to 50%. It is generally recognized that the inhibition of AChE and ChE activities in human blood by 75% is an indicator of danger and requires taking immediate measures to discontinue the exposure.

Since OPC block ChE selectively in every cholinergic structure (M- and N-cholinoreceptive systems), practically all physiological systems and organs can be involved in a pathological process. At that, changes in activity of the central and peripheral nervous systems as well as the resulting impairments in respiratory and cardiac activities have a critical influence on an outcome of the poisoning.

A number of OPC are capable of rendering a delayed neurotoxic effect (DNE). This effect becomes apparent gradually, after a certain latent period (usually 14 to 21 days, sometimes 1 to 5 years after the acute poisoning survived) and is characterized clinically by the development of ataxia, muscular weakness, paresis and paralysis of the extremities. Morphologically, it is characterized by fiber demyelinization of spinal pathways and peripheral nerves. Till present time, near 40,000 cases have been described, when paresis and paralysis developed in human beings as a result of their exposure to OPC (TOCP, mipaphox, chloropyrophos, trichlorfon, etc.) [1].

The effective theraphy for prevention on DNE is unknown. The forecast of OPC caused delayed polyneuropathy is one of the main task of modern medicine.

DNE mechanism for OPC has not been elucidated finally. Neither in experiment, nor in clinical material, a direct dependence has been established between their anti-cholinesterase and neuroparalytic effects. The phosphorylation of a protein, which belongs to carboxyesterases and is named neurotoxic (or neuropathy) esterase (NTE), is considered to be an important pathogenetic component of injury [2]. Delayed neuropathies develop only in the case of exposure to such OPC, which inhibit NTE by 70 to 80%, with the DNE development being connected not only with NTE inhibition, but also with its subsequent aging. The aging of NTE that was inhibited by OPC (TOCP, leptophos, mipaphox, aphos, etc.) occurs extremely rapidly, as early as 1 to 24 hours following a single exposure.

We examined in vivo and in vitro more than 30 OPC (phosphates, phosphonates and etc.) with different structures [3, 4].

It was shown that the first biochemical lesion in the nervous system after OP treatment is significantly depressing of NE in nervous tissue.

It was established in vivo and in vitro experiments on hens, guinea-pigs and rats that in the pathogenesis of developing OPIDP, induced by TOCP, aphos, oxyphosphonate, leptophos, the leading role belonged to inhibition (on 70-90%) of brain NE, which took place in the first 24 hours after OPC administration in toxic doses (LD_{50}). OPC such as ortek, ofunac, malathion, etaphos etc., which have low inhibitory activity to NE (inhibition on 3-30 %), did not produce clinical disorders in hens.

The histological investigation showed that these OPC did not produce degeneration and demyelinization in hen's spinal cord and peripheral nervous system.

Correlation between NE inhibition in hen, guinea-pig and rat brain such as in lymphocytes from peripheral human blood has been found. It was concluded that NE activity may be used for delayed polyneuropathy screening of OPC.

Antidotic agents presently used in poisoning with organophosphorus substances are aimed at blocking the cholinergic effects resulted from a rise in ACh level as well as lowering its concentration in synapses.

A principle of the therapy for acute poisonings with anti-cholinesterase compounds (OPC, carbamates) lies in the complex performance of specific antidotic therapy including methods for poison excretion and intensive resuscitation measures.

The specific therapy consists in concurrent use of two antidotes differing by mechanism of action: cholinolytics eliminating anti-cholinesterase effects on CNS, and cholinesterase reactivators (ChR) ensuring the restoration of inhibited enzymatic activity.

A reduction of the mediator concentration in synapses can be also achieved by the AChE activation, inhibition of mediator release into synaptic slit, slowing-down of reverse choline intake, as well as by using AChE compounds [5].

For the purpose of therapy, atropine is used in high doses and repeatedly, since its action ceases much sooner than OPC effects.

At the same time, with the accumulation of both experimental and clinical data, it becomes evident that atropine and other cholinolytic compounds only exert a pronounced therapeutic effect in poisonings of mild and moderate severity, and are less efficacious in severe intoxications.

The essence of ChR action consists in dephosphorylation of the enzyme inhibited, which becomes apparent as restoration of its activity, i.e. ability to perform enzymatic hydrolysis. Therapeutic efficacy of ChR is associated with their capability to eliminate toxic effects of anti-cholinesterases on nicotine receptors.

Based on studying intimate components of the OPC and ChR mechanisms of action, it was established that summary antidotic effect of oxime, along with reactivation of phosphorylated AChE, can consist of the following pharmacological effects: deblockade of neuromuscular conduction (ability to destroy the OPC+cholinoreceptor complex), ganglioblocking action, M- and N-cholinolytic activity, AChE protection against irreversible OPC inhibition , inhibition of the AChE aging reaction, direct chemical interaction with ACh in blood.

The analysis of literature allows several principal directions to be singled out for the development of efficacious ChR [6].

Referred to ChR of the *first generation* (the development started in the middle 60-ies of the last century) are pralidoxime (2-PAM-iodide and chloride), dipiroxime (TMB-4), toxogonine (obidoxime, LuH-6), P2S, isonitrosine. They are means of medical protection in cases of OPC poisoning and are implemented as antidotes in medical practice.

The *first generation* ChR are well studied, possess high specific activity in the animal treatment for affection with sarin and substances of VX type [7].

[At the same time, being quaternary compounds, they penetrate poorly biological membranes and so have predominantly peripheral type of action.

A significant disadvantage of these compounds lies in their relatively high toxicity (DL_{50} = 100-220 mg/kg, i.m., mice); some of them (dipiroxime) are able to form difficult-to-hydrolyze esters with OPC, whose toxicity is higher than that of the initial compounds. Severe hepatic lesions, unusual for OPC toxic effects, were described, when dipiroxime was used for a long time.

Another disadvantage of the first generation reactivators lies in their lacking of antidotic effect in respect to affections induced by soman, which causes fast aging of the phosphorylated enzyme.

The end 70-ies was noted by the development of universal efficacious agents for the therapy of OPC affection, H-oximes or reactivators of the second generation, by means of structural modification of the first generation reactivators. The most efficacious compounds were synthesized in Germany under the direction of I. Hagedorn [8].

Among H-oximes, high activity is demonstrated by such compounds, as HI-6, HS-6, HGG-12, HGG-42, Hlo-7.

HI-6 is the most efficacious antidote in poisonings with soman. This agent has less toxicity as compared with HS, H6-6, toxogonine and 2-PAM.

Because of their weak lipophily, H-oximes overcome poorly the hematoencephalic barrier. Their antidotic action is based on reactivation of peripheral AChE (in blood and respiratory musculature).

A positive property of the most active reactivators of the second generation lies in their ability to reduce 2 to 2.5 times aging rate of AChE inhibited with soman .

Efficacy of oximes grows up, when they are combined with atropine sulfate. HI-6 reserves for itself superiority over other H-oximes.

H-oximes used in combination with atropine sulfate against a background of the intramuscular or inhalation administration of soman are able to protect animals (dogs, monkeys) from 4-5 DL_{50} of this poison [9].

A significant disadvantage of the 1st and 2nd generation ChR created on the basis of pyridine aldoximes is their own, relatively high toxicity (DL_{50} amounts to 146-250 mg/kg, i.m.) as well as low stability in aqueous solutions.

HI-6 of Bulgarian production has proved its low toxicity and high efficacy [10].

In different years, Russian toxicologists implemented into practice antidotic agents for medical aid to those affected with high toxic organo-phosphorus poisonous substance (PS), such as *aphin, dipiroxime, budaxim.* The studies have been completed on development and implementation into industrial production of an antidote for self-care and mutual aid, *pelixime*, as well as a ChR, *carboxime* (1-4-methyl-5-[2^1-(benzyl dimethyl ammonium) ethyl]) carbamoilpyridinium-2-aldoxime dichloride [11].

The antidotic complex of atropine (0.1% solution of atropine sulfate in 1.0-ml ampoules) and carboxime (15% injection solution in 1.0-ml ampoules) is proposed as a basic antidotic means aimed at the usage for the OPC destruction.

High therapeutic effect was received with the usage of atropine (10 mg/kg, i.m.) and carboxime (30 mg/kg, i.m.) in animals poisoned with sarin, soman and VX.

Being not inferior to HI-6 in efficacy, carboxime is more convenient, since it is proposed as a ready-for-use solution for injection.

The usage of atropine in combination with carboxime and caventon prevents the development of vascular disorders (normalizes state of cerebral circulation) in animals.

Convulsions are among the most severe symptoms of intoxication with high toxic OPC.

For this, antidotic formulas are supplemented with anticonvulsants of different chemical structures (diazepam, clonazepam, phenazepam, etc.).

Quaternary ChR, which possesses lipophily at the expense of long allyl radicals present in their molecule, seem to be of certain interest. A representative of this type compounds is alloxime (an analogue of 2-PAM) synthesized and studied at Institute of Pharmacology and Toxicology (Kyiv, Ukraine) [12].

The compound gave good account of oneself in clinic in the therapy for acute OPP poisonings.

In contrast to dipiroxime, alloxime in therapeutic doses demonstrates a pronounced therapeutic effect in poisonings induced both by sarin and soman (rats and dogs) and by high toxic carbamates (carbofuran, ellocron, pirimor).

In Ukraine during the 70-80-ies of the last century, a new direction appeared in the development of ChR on the basis of previously unknown thiohydroxime esters, which are close structurally to acetylcholine [13-15].

When working out on this direction, ***diethyxime:***

S-[2-(Diethylamino) ethyl]4-Bromobenzothiohydroximate has been deveoped, studied in detail and implemented into medical practice.

This is a low toxic ChE reactivator (DL_{50} = 950 mg/kg, rats, i.m.) of central action.

In acute poisoning with OPC or carbamate pesticides, percent of AChE reactivation in different parts of the brain amounts to 60-80%.

The mechanism of diethyxime therapeutic action on the neuromuscular conduction in poisonings with anti-cholinesterase compounds is associated both with reactivation of synaptic AChE and a deblockaded postsynaptic membrane cholinoreceptor and with normalization of mediator release by the postsynaptic membrane. In case of the combined usage of diethyxime with TMB-4 and atropine sulfate (each in a dose of 10 mg/kg, i.m.), the effect of their antidotic action potentiation is observed. A value of IP amounts to 15 in acute poisoning of rats with DDVP, which was confirmed by Indian scientists. Unfortunately, the therapeutic efficacy of diethyxime has been studied insufficiently on the model of intoxication with sarin and Vx. A subsequent development of this direction became the purposeful designing of reativator-cholinolytics in the rank of thiohydroxime esters. We have patented a number of designed compounds [16,17]. It was synthesized over 30 compounds of similar structure, including thiohydroxime analogues of cholinolytics (*amisile, thiophene, dipraphene, arpenal)*.

AChE of high purity and peptide toxins isolated from the venom of cobra Naja naja oxiana are efficacious as prophylactic means in poisonings with organophosphorus inhibitors of ChE [18]. Reversible ChE inhibitors of the pyridostigmine type are used for prophylaxis of OPC poisonings. Good effect was gained in the combined use of pyridostigmine and diazepam [19]. Unortunately there is no production of ChE-reactivators in Ukraine at present time.

A principle of the therapy for acute poisonings with OPC lies in the complex performance of cholinolytics (atropine, scopolamine, aprofene, amisile etc.) with reversible ChE inhibitors (*pyridostigmine, proserine* etc.) or cholinolytics with tranquillizers (*diazepam, fenozepame etc*.)

The development of efficient means for prevention of neurotoxic action of OPC (TOCP, mipaphox, dichlorvos, DFP, etc.) of the delayed type (DNE) remains to be relevant.

The usage of specific therapeutic means (quaternary ChE reactivators, cholinolytics) with this purpose prevents only the development of cholinergic symptoms of intoxication, but does not exert influence on the development of paresis and paralysis in the remote period.

In the experiment (rats and chickens), we showed that central ChR (diethyxime and other synthesized thiohydroxime esters – LA-81) were capable of reactivating activity of NTE in the brain of chickens and rats (by 56-71%) poisoned with certain OPC (triorthocresylphosphate, leptophos), thereby attenuating the development of neuropathies in the remote period. One cannot also exclude a possibility of direct interaction between ChR and OPC resulting in neutralization of the latter. The best effect was obtained in case of the prophylactic-and-treatment usage of oximes and an inductor of the hepatic monooxygenase system, e.g. phenobarbital (20 mg/kg, intragastrally). The electrophysiological and morphological studies confirmed positive effects of the therapeutic agents indicated on neural structures and skeletal musculature [20].

At present, professional toxicologists should determine more clearly a range of neurotoxicants, which can be used with maximal probability in terrorist acts, and concentrate their efforts on the development of combined antidotic means composed of ChR and cholinolytics (or agents possessing concurrently reactivating and cholinolytic activities), anticonvulsants and reversible AChE inhibitors.

At the same time, while working out such combinations, one should solve a number of problems - compatibility, solubility, and stability of various ingredients in a combination.

In view of possible use in terrorist acts of neuroparalytic OPC, the development of antidotic-and-therapeutic means is an extreme relevant problem.

REFERENCES

1. (Makhaeva G.F., Malygin V.V.) Махаева Г.Ф., Малыгин В.В //Агрохимия.-1987.-№12.- С.103-124.
2. Johnson M.K.//Arch.Toxicol.-1977.-№3.-P.113-115.
3. (Kagan Y. S., Kokshareva N.V., Tkachenko I.I.) Каган Ю.С., Кокшарева Н.В., Ткаченко И.И. //Бюл. эксп. биол. и мед. -1986.- № 9. - С. 310-312.
4. Tkachenko I.I., Kokshareva N.V., Kagan Yu.S. et al. //Fresenius Env. Bul. - 1993. - N 2. - P. 131-1362.
5. Daris F.F.// Gov.Rep. Ann. Index.-1988.-V.88.-№19.-P.165.
6. (Mokhort N.A., Pritula T.P.) Мохорт Н.А., Притула Т.П.// Современные проблемы токсикологии.- 2003.- №2.- С.18-26
7. Krummer S., Thiermann., Worec F.//Arch.Toxicol.-2002.-V.76.- №10.-P.589-595.
8. Hagedorn I., Stark I., Lorenz H.P. // Angew. Chem. —1972.-V. 11, N4. - P. 307-309.
9. Clement I.W., Lockwood P.A. //Toxicol.Appl.Pharmacol.-1982.-V.64.- N1.- P.140-146.
10. Dishovsky Ch. // Abstract of the 1-st Congress of Ukrainian Toxicologists – 2001.-P.-49-50.
11. (Petrov A.N., Netchiporenko S.N.) Петров А.Н. Нечипоренко С.П. // В кн.: Медико-биологические проблемы противолучевой и противохимической защиты.- Санкт-Петербург.- «Издательство Фолиант».- 2004. – С.367-368.
12. (Loboda Y.I.) Лобода Ю.И.// Фармакол. и токсикол.- 1990.-т. 53.-№1.- С. 24-28.
13. (Krivencuk V.E., Petrunkin V.E.) Кривенчук В.Е., Петрунькин В.Е.//А.с. 287931(СССР).- Б.И. - 1970.- №36.-С.27.
14. (Krivencuk V.E.) Кривенчук В.Е.//А.с.683744(СССР).- Б.И. - 1979.- №33.- С.23.
15. (Krivencuk V.E, Petrunkin V.E.) Кривенчук В.Е., Петрунькин В.Е. // А.с. 419523 (СССР).-Б.И.-1974.-№10.
17. (Krivencuk V.E., Kokshareva N. V.) Кривенчук В.Е., Кокшарева Н.В.//А.с. 579761 (СССР) (не публикуется).
18. (Anikienko K.A., Dobrianskiy V.S.) Аникиенко К.А., Добрянский В.С.// В кн.: Медико-биологические проблемы противолучевой и противохимической защиты.- Санкт-Петербург.- «Издательство Фолиант».- 2004. – С.277.
19. Drakontides A., Baker T., // Noxicol. Appl. Pharmacol.-1983.-V.70. -N 2.-P.-411-422.
20. Kokshareva N.V., Tkachenko I.I., Krivenchuk V.E. //Toxicology Letters.-(EUROTOX-96.- Alicante, Spain),-1996.-P.2

Chapter 12

CERTAIN PROBLEMS OF RENDERING MEDICAL ASSISTANCE UNDER ACTS OF CHEMICAL TERRORISM

S. I. Khmel', Yu. P. Litvin, A.N. Gninenko
Dnipropetrovsk State Medical Academy, Dnepropetrovsk, Ukraine

Keywords: chemical tertrorism; diagnostics; exogenic intoxication; extreme situations

Today there are no means to predict possible terrorist acts in a reliable way , and information (apart from intelligence data) allowing identification of their forms and methods, is not available.

Owing to the high accessibility of firearms and explosives, most of the terrorist acts committed so far were efficient, and to a greater extent meant to frighten the citizens (Israel) or to imposeon autorities (Iraq) decisions beneficial to terrorists .. To a lesser extent, they were intended for causing considerable damage to property or mass destruction of people.

In this connection, considerable interest is to be given to the acts of terrorism utilizing “alternative” methods of action, like those stated below:

- using airborne vehicles (both piloted and pilotless ones) as a tool for striking housing and industrial objects including nuclear power plants and chemical enterprises;
- using explosive devices for destruction of objects where toxicants are stored;
- intentional disruption of production process creating conditions for technogenic accidents and disasters.

There are over 1800 industrial enterprises operating in Ukraine. They are manufacturing, using and storing nearly 280 thousand tons of toxicants including 9,8 thousand tons of chlorine and 178 thousand tons of ammonia. An extensive network of gas- and oil pipelines and interstate ammonia-line is available. Potential victims in these areas can be about 20 million persons (38,5 % of citizens residing in this country). Usage of products manufactured and stored at these enterprises as a tool of chemical terrorism, both in Ukraine and in the other countries, is not excluded.

C. Dishovsky et al. (eds.), Medical Treatment of Intoxications and Decontamination of Chemical Agents in the Area of Terrorist Attack, 109–112.

It is promoted by low prices for toxicants production, easy storage and bringing to operational readiness, secrecy of delivery by any means of transport to the place of terrorist act. The act of terrorism is committed in secret. It results in sizeable mass casualties, occurring instantly or having prolonged action, such as severe psychic disorders in people, causing disorganization of authorities and power structures, and necessitates rendering efficient medical assistance.

Means for arrangement of terrorist acts can be stolen during storage or transportation, or manufactured by chemists using rather simple laboratory equipment and a usual exhaust hood.

Preference is given to highly toxic chemical agents with minimum organoleptic characteristics, which are introduced through inhalation, featuring instant or retarded effect. Volatility of the substance promoting fast spreading, and maximum concentration of its vapors in the air which can greatly exceed lethal concentration, are of ultimate importance.

Optimal toxic substances are military agents both registered for combat use and unregistered but readily available in production (phosgene, prussic acid). Places and methods of toxic agents' use are deliberately chosen; anyway, public places with intensive ventilation are preferable (subways, railway stations, supermarkets, sports centers, discotheques).

Taking into consideration the peculiar features of chemical terrorist acts, namely:

- quick or retarded spread of damage;
- non-availability or inefficiency of protective means;
- lack of data on the toxicant or its tactical blend prior to indication;
- possibility of affecting many people in confined space and, accordingly, simultaneity and suddenness of damage by fast-acting toxicants, or occurrence of damage in various places of the inhabited locality at once when using toxicants of retarded action;
- occurrence of severe incidents threatening life of injured persons, and requiring specific therapy or maintenance of vital functions of large quantities of people at the same time;
- lack of properly trained experts among physicians, who render assistance, and accordingly, erroneous interpretation of symptoms and inadequate medical treatment;
- difficulties in predicting the course of events because of the contaminated areas appearance and prolonged affection of people in case of using persistent toxic substances;
- non-readiness of the local public health authorities to mass arrival of affected people, non-availability of sufficient amount of materials and medical resources, in particular, during the first hours upon toxicant effect; lack of knowledge on rendering medical assistance in the event of mass damage;
- impossibility of quick and reliable indication if the substances used differ from the registered toxicants, or are applied as tactical blends (persistent with non-persistent ones, etc.);

- non-availability of required facilities and means for degassing contaminated objects;

- dependence of efficient work of governmental authorities and medical institutions on the degree and scope of toxic injuries, chemical properties and quantity of toxicants used, place and time of their usage, promoting microclimatic conditions, quantity of people in effective area, occurrence of panic, psychoses, hallucinations, disturbances, which can disorganize and extremely impede remedial operations.

Therefore, governmental authorities and local public health bodies need to apply great material, technical and physical efforts on taking a number of special measures to eliminate consequences of the act of chemical terrorism.

Firstly, it concerns the proper protection of persons taking part in maintaining the law order, rendering medical assistance, conveying injured persons and degassing contaminated objects, as well as personnel of hospitals where the affected persons are brought.

An essential role is given to provision with medical protection means (individual degassing packs, antidotes, gas-masks, skin remedies) for all exposed people, since their shortage often results in troubles, panic, affective manifestations, etc.

Timely rendering of medical assistance and evacuation of people can depend on carrying capacity of escalators, exits, condition of roads and other ways of evacuation, availability of proper evacuation means, and evacuation periods.

When arranging proper medical care, one can' help admitting the ultimate importance of medical sorting, performed in compliance with the conventional scheme taking into account the factors below:

- threat to the people around;
- degree of medical assistance required;
- need in evacuation and its order.

The second and the third factors are the most critical; they can stipulate breaking into groups below.

The first group: persons in grave condition threatening their life (acute respiratory failure and cardiovascular collapse, coma, convulsions, paralytic manifestations). Upon first aid provided, these persons are to be evacuated to specialized medical institutions as soon as possible, preferably by ambulance cars, equipped with reanimation facilities.

The second group: persons with clinical symptoms of intoxication, but not requiring the emergency care by their life-saving indications. Upon first aid provided, the following medical treatment can be put off till arrival to the hospital.

The third group: persons with light degree of intoxication, and people exposed to toxicant action but displaying no clinical symptoms. These people need observation for at least one day upon first aid rendering, and subsequent outpatient treatment and doctor supervision is to be provided.

In the area affected, medical care is rendered in maximum possible amount and in the shortest time, both by rescue teams and first-aid brigades, promoting favorable outcome of the accident.

Rendering of medical assistance can be difficult because of lack of due methods and procedures allowing assessing probable sanitary losses under the acts of chemical terrorism, and non-availability of adequate staff of toxicologists and toxicological centers in most of the regions in the country. Therefore, on the one hand, due normative documents regulating the operations of governmental authorities in the event of the acts of chemical terrorism are to be drawn up. On the other hand, it is necessary to form a proper reserve of medical aids and protective means; to provide for adequate training of medical experts, first-aid brigades and rescuers in algorithms of operations in the event of terrorist acts; to arrange mobile medical teams for rendering first aid and qualified medical assistance in the areas of possible terrorist acts. All above-listed measures require task-oriented actions at the governmental level.

Chapter 13

EFFICACY OF PRETREATMENT AND TREATMENT AGAINST SOMAN INTOXICATION

Ingrid H.C.H.M. Philippens[1,2], Marjan J. Jongsma[1], and Raymond A.P. Vanwersch[1]

[1]TNO Prins Maurits Lab (TNO-PML), Research Group Medical Countermeasures, P.O.Box 45, 2280 AA Rijswijk ZH, The Netherlands.

PsychoPharmacology, Utrecht Institute Pharmaceutical Sciences, University Utrecht.

Summary: The efficacy against lethality and post-intoxication incapacitation after 2x LD50 soman of different subacute pretreatment scenarios of 12 days was tested with or without post-intoxication therapy in guinea pigs. These pretreatment regimes were 1) the currently used pyridostigmine (PYR, 0.04 mg/kg/hr), 2) the combination of physostigmine (PHY, 0.025 mg/kg/hr) with the muscarinic receptor antagonist scopolamine (SCO, 0.018 mg/kg/hr), and 3) the combination of PHY with the anti-Parkinson drug procyclidine (PC, 3 mg/kg, sc). The post-intoxication therapy consisted of HI-6 (21.4 mg/kg, im), atropine sulphate (AS, 0.085 mg/kg, im), and diazepam (DZP, 0.21 mg/kg, im). Behavioral and observational read-out systems were used to elucidate the severity of soman induced incapacitation.

There were no large differences in the symptomatology between animals pretreated with PHY+SCO or with PYR. On the other hand, animals pretreated with PHY+PC did not show convulsive activity as was found in the animals treated with the other scenarios. In some cases the post-intoxication therapy was even not necessary.

Although the symptomatology between PHY+SCO and PYR were comparable the effects on incapacitation and survival were not comparable: all animals pretreated with PYR and the post- intoxication therapy did not perform in the behavioral test systems and died within 24 hours. This was the same result as what was found in animals treated only with the post-intoxication therapy. The mortality in animals pretreated with PHY+SCO was 25%. The addition of a post intoxication therapy in PHY+SCO pretreated animals did not improve the efficacy. On the other hand, animals pretreated with PHY+PC all survived even without post-intoxication therapy. The behavioural performance was similar for animals pretreated with PHY+SCO or PHY+PC.

The combination of PHY and SCO seems to be a good alternative for the current PYR pretreatment, in particular since this combination showed no side effects in previous studies with guinea pigs and marmoset monkeys and protects efficiently against 2x LD_{50} soman. The addition of PC to PHY, instead of SCO, augments the efficacy of the pretreatment against soman poisoning. In that case an additional post-intoxication therapy is

C. Dishovsky et al. (eds.), Medical Treatment of Intoxications and Decontamination of Chemical Agents in the Area of Terrorist Attack, 113–121.

not necessary. The combination of PHY + PC seems to be very promising and should be further investigated.

Keywords: behaviour; guinea pig; physostigmine, procyclidine; pretreatment; physostigmine; scopolamine; soman

1. INTRODUCTION

Exposure to nerve agents is not restricted to the battlefield. Possible terrorist use of these weapons and the destruction of the chemical weapon depots certainly will increase the risk of exposure. Since treatment for intoxication with at least some of these organophosphorus (OP) acetyl-cholinesterase (AChE) inhibitors is still far from ideal, research efforts are devoted towards finding an effective treatment.

Currently PYR is used in most NATO countries as a prophylaxis against OP intoxication in combination with a therapy consisting of an oxime, AS, and an anticonvulsant DZP. PYR with a post intoxication therapy protects effectively against lethality in a number of species (Gordon, 1978; Leadbeater, 1985; Dirnhuber *et al.*, 1979). However, PYR alone doesn't protect against the toxicant (Gordon *et al.*, 1978). The limited protection of PYR is mainly caused by its lag to enter the brain because of its quaternary nitrogen atom. A single i.m. injection of PYR (131 μg/kg) caused 58.5 % inhibition of blood AChE and no inhibition of brain AChE (Ray, 1991). This may lead to brain damage and post-intoxication incapacitation, which is in accordance with findings on behavior: no central effects were found after PYR administration compared with low dose levels of PHY or soman (Wolthuis *et al.*, 1995). Therefore, PYR may not protect the central nerve system. A promising alternative for PYR is PHY, which does enter the brain easily. To prevent central side effects, PHY will be combined with the muscarinic antagonist SCO. There are also assumptions that other systems, like the NMDA receptor complex, may play a role in the protection against OP intoxication (Raveh *et al.,* 1999). In this study the efficacy of prophylaxis with PYR against soman intoxication was compared with PHY combined with SCO or with the NMDA antagonist PC. This was performed in combination with a post-intoxication therapy. Efficacy against soman was tested on mortality, symptomatology and incapacitation. Most effects can be expected to be centrally mediated effects that may induce changes in behavior. Therefore, behavioral read-out systems were used to elucidate the severity of soman induced incapacitation.

2. METHODS

Animals: Male Dunkin-Hartley albino guinea pigs (Harlan) (350-400 g) were used. The ambient temperature was regulated between 20-22°C. Relative humidity was kept over 50%. Animals had free access to food and water. The experiments received approval by an ethical committee.

Drug solutions and implantation of osmotic mini-pumps: Physostigmine hemisulphate and procyclidine hydrochlorid were obtained from Sigma (St.Louis, U.S.A.), scopolamine hydrobromid from Merck (Darmstadt, Germany), atropine sulphate was obtained from ACF (Amsterdam, The Netherlands), and diazepam from Roche (The Netherlands). HI-6 was made available by the Defence Research Establishment, Suffield, Canada. Soman (O-pinacolyl methylphosphonofluoridate) was synthesised at TNO. Alzet® Osmotic Mini-pumps with a constant delivery rate of 0.55 µl/hr (Model 2002, Alza Corp., Palo Alto, USA) were used to deliver PYR, PHY and SCO. The vehicle consisted of 20% propylene glycol, 10% ethanol and 70% water. The pumps were implanted subcutaneously under isoflurane/O_2 inhalation anesthesia.

Table 1. Overview of the different treatment groups

group	N	pretreatment	intoxication	Post-intoxication therapy	Administration scheme of therapy
1	8	PHY +SCO	2x LD50 Soman	-	-
2	8	PHY +SCO	2x LD50 Soman	HI-6 + AS + DZP	After one minute
3	8	PHY +SCO	2x LD50 Soman	HI-6 + AS + DZP	After first symptoms appeared
4	8	PHY + PC	2x LD50 Soman	HI-6 + AS + DZP	After first symptoms appeared
5	8	PYR	2x LD50 Soman	HI-6 + AS + DZP	After first symptoms appeared
6	5	-	2x LD50 Soman	HI-6 + AS + DZP	After first symptoms appeared
7	5	-	2x LD50 Soman	-	-

PHY 0.025 mg/kg/hr; SCO 0.018 mg/kg/hr; PYR 0.04 mg/kg/hr; PC 3 mg/kg, sc; HI-6 21.4 mg/kg, im; AS: atropine 0.085 mg/kg, im; DZP: diazepam 0.21 mg/kg, im.

Study design: The efficacy against soman of different subacute pretreatment scenarios was tested with or without post-intoxication therapy in guinea pigs (see Table 1). After training of the shuttle box task, some animals received an osmotic mini-pump containing either PHY with SCO (group 1, 2, 3), or PHY alone (group 4), or PYR (group 5). PC was given 30 min prior to soman intoxication (group 4). Twelve days after pump insertion all guinea pigs were intoxicated with 2x LD_{50} soman sc (48 ug/kg). Animals from group 2 received the post-intoxication therapy one minute after soman intoxication. Animals from groups 3, 4, 5, 6 received the post-intoxication therapy after the first symptoms appeared. After intoxication the animals were observed for clinical signs. After most signs had disappeared the animals were tested for incapacitation. This was repeated one day and one week later. Blood-samples were taken for the determination of AChE activity before and after the pretreatment period and one hour after soman intoxication.

Behavior: Shuttle box: The performance of the active avoidance was measured in a two-way shuttle box, consisting of two equal compartments, connected by a photo-cell-guarded gate (Philippens *et al.*, 1992). Guinea pigs have to learn how to avoid a stream of air (6 l/s, air tube diameter 1 cm) within 10 seconds after presentation of a sound stimulus with a random inter-trial interval (20-30 s). The number of correct avoidance responses (CARs) was used to express the retrieval of learned behavior. Auditory Startle Response: The startle response of 200 ms duration was measured by a transducer connected with the platform on which the animals are resting, registering the force exerted by the animal upon presentation of the startle signal (120 dB, 10 kHz, 20 ms). The amplitude of the startle response was used to express the motor reaction of the startle reflex (Philippens *et al.*, 1997). Observation of symptoms: Signs as chewing, hyper-salivation, trembling, tremors, convulsions, and dyspneoa were observed. The occurrence of each symptom was expressed as a percentage of the total observations. The number of animals suffering from each symptom was used to indicate the severity of the intoxication symptoms. *Blood AChE activity:* This was assessed using a radiometric method (Johnson and Russell, 1975). Blood samples (25 μl) were mixed with 1% saponin. The ACh end-concentration used was 12 mM; [^{3}H]ACh iodide (NEN, Germany) was diluted to an activity of 602 $MBq.mmol^{-1}$. Electric eel AChE was used as reference.

Statistical analysis: For statistical analysis of the behavioral tests an analysis of variance (two-way ANOVA) was used. For the symptomatology a Fisher exact probability test or an unpaired t-test with Welch's correction was used. In all tests p values <0.05 were considered significant.

3. RESULTS

Blood AChE-inhibition: The mean blood AChE-inhibition after pretreatment, measured as a percentage of the baseline value, was 20-35 % (see Table 2). The AChE-inhibition measured one hour after soman followed by a post-intoxication therapy, was significantly lower in all groups (t-test with Welch's correction, $p < 0.05$). The inhibition of AChE after soman was less in animals pretreated with PYR or with PHY combined with PC (resp. 85.5 ± 1 % and 89.4 ± 2 %).

Efficacy against soman induced lethality: In Table 2 the survival after soman is shown. All untreated guinea pigs died within one hour. Unpretreated and PYR pretreated animals died within 24-hours. In the PHY+SCO group without an additional therapy only one animal died shortly after soman. Remarkably, pretreatment with PHY+PC protected completely against lethality. In 3 cases an additional post-intoxication therapy was even not necessary. All animals that survived 24 hours after soman intoxication stayed alive during the complete study. The effects on survival after pretreatment with PHY combined with SCO or PC was found to be significantly higher compared with PYR or no pretreatment (Fisher exact probability test, $p<0.05$). There were no differences in lethality between PYR and no pretreatment (Fisher exact probability test, $p = 1$).

Table 2. The Blood AChE activity after pretreatment and 1 h after 2xLD_{50} soman and survival.

group	N	Pre-treatment	HI-6+AS+DZP therapy	AChE act. after pretreatment % (SEM)	AChE act. 1h after soman % (SEM)	Survival after 2xLD50 soman 1 h	12 h	24 h
1	8	PHY +SCO	-	63.4 (5.9)	6.4 (1.2)	7/8	7/8	7/8
2	8	PHY +SCO	After 1 minute	76.3 (6.6)	6.6 (1.2)			
3	8	PHY +SCO	On indication	74.6 (8.4)	8.1 (2.0)	8/8	7/8	6/8
4	8	PHY + PC	On indication	70.2 (8.3)	10.6 (1.8)	8/8	8/8	8/8
5	8	PYR	On indication	79.1 (9.3)	14.5 (1.1)	7/8	2/8	0/8
6	5	-	On indication	NT	NT	5/5	1/5	0/5
7	5	-	-	NT	NT	0/5	-	-

Survival is expressed as the number of animals alive / group size. NT: not tested.

Post-intoxication symptoms: The symptomatology after 2x LD_{50} soman followed by a post-intoxication therapy of animals pretreated with PYR was comparable with no pretreatment (see Fig. 1). Pretreatment with PHY+SCO improved the signs compared with PYR. This was mainly found on severe symptoms. On the other hand, animals pretreated with PHY+PC did not show any worse symptoms as was found in the other groups. In some cases the post-intoxication therapy was not necessary because no symptoms were present.

Post-intoxication incapacitation: The post-intoxication incapacitation was measured immediately after most symptoms disappeared, 24 hours, and one week after the intoxication. However, one day after soman intoxication no animals from the PYR group were available. Since the un-pretreated animals all died very shortly after soman no post-intoxication incapacitation was tested in these animals. The shuttle-box performance declined after soman intoxication with approximately 60% in all animals pretreated with PHY and SCO or PC. This decrease of performance was significant in the groups treated with a post-intoxication therapy (not

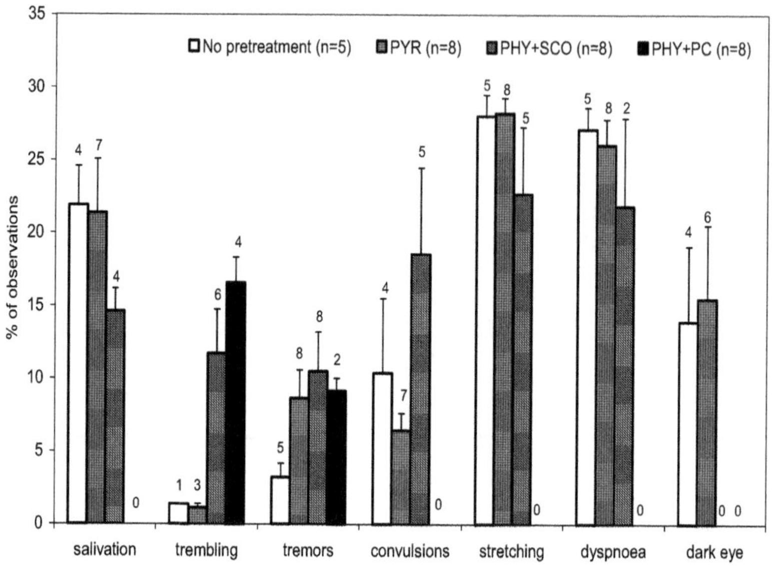

Figure 1. Mean (SEM) severity of symptoms after soman followed by a post-intoxication therapy on indication. The occurrence of symptoms is expressed as a percentage of observations. The number of animals in which the symptoms were observed is indicated above the bars.

shown) (ANOVA: $p < 0.05$). The decline of the performance was not significant between these treatment groups. One day later these animals recovered to baseline. Animals pretreated with PYR did not perform at all in this task after soman (CARs=0). After soman intoxication the auditory startle reflex of the guinea pigs pretreated with PYR was almost absent: no startle reflex was observed (not shown). On the other hand, there was a tendency towards an increase of the amplitude in animals pretreated with PHY+SCO and PHY+PC combined with a post-intoxication therapy, with its maximum one day after soman. PHY+SCO pretreated animals without post-intoxication therapy, performed normal in this task.

4. DISCUSSION

The efficacy of the PYR against soman intoxication was compared with PHY combined with SCO or with PC combined with a post-intoxication therapy. The efficacy was tested on mortality, symptomatology, and incapacitation. There were no big differences in the symptomatology between animals pretreated with PHY+SCO or with PYR. On the other hand, animals pretreated with PHY+PC did not show convulsive activity as was found in the other groups. In some cases the post-intoxication therapy was even not necessary. Although the symptomatology between the animals treated with PHY+SCO and PYR were very similar the survival was not comparable: all animals treated with PYR and the post-intoxication therapy died within 24 hours. This was the same result as what was found in unpretreated animals receiving only the post-intoxication therapy. The mortality in the guinea pigs pretreated with PHY+SCO was resp. 1/8, 3/8, 2/8. The protection with PHY+SCO could not improve further by a post-intoxication therapy.

Pretreatment with PYR did also not prevent the animals from post-intoxication incapacitation. There was no response of the startle reflex and these animals did not perform in the shuttle box task. Animals pretreated with PHY+SCO or PHY+PC were in a much better condition concerning incapacitation. Remarkably, the addition of a post-intoxication therapy did also not improve the efficacy on incapacitation in animals pretreated with PHY+SCO.

It is assumed that the protection of carbamates like PYR and PHY is based on the reversible inhibition of the enzyme AChE. However, from PHY it is known that it can affect receptors in a direct manner (Albuquerque *et al.*, 1984; Albuquerque *et al.*, 1988; Sherby *et al.*, 1984, Van den Beukel *et al.*, 1998). The ED_{50} of PHYs' agonism at the nicotinergic receptor even appears to be lower than its IC_{50} of AChE inhibition (Albuquerque *et al.*, 1988). Interestingly, nerve agents VX and soman also affect ACh receptors (Bakry *et al.*, 1988). They reported that soman can act as a partial agonist of the nicotinergic ACh receptor. VX and soman may also affect a small population of muscarinergic receptors. This observation suggested that the toxicity of soman might involve an additional action on cholinergic receptors. This could explain the fact that PHY is more effective in the protection against soman intoxication than PYR. Interesting results were found with the optical isomer of PHY, (+)-PHY that hardly inhibit ChE. This isomer did not inhibit and therefore did not protect AChE, whereas the symptomatology and the post-intoxication incapacitation after 2x LD_{50} soman were comparable with the regularly used (-)-PHY (Philippens *et al.*, 2003).

Surprisingly, animals pretreated with PHY+PC all survived even without post-intoxication therapy. This effect is presumably based on the fact that progressive seizure activity after soman intoxication recruited other neurotransmittersystems like the NMDA system. Indeed, PC, a drug with anticholinergic and anti-NMDA activity, combined with diazepam and pentobarbital successfully terminated soman-induced seizures in rats (Myhrer et al., 2003). PC in combination with PHY and a post-intoxication therapy with atropine sulphate and HI-6 offer a very high protection ratio of 21.5 in guinea pigs against soman (Kim *et al.,* 2002).

It can be concluded that pretreated with PYR did not protect against lethality and post-intoxication incapacitation after soman. Prophylaxis with PHY+SCO seems to be a good alternative for PYR. In particular since PHY+SCO pretreatment showed no side effects in previous studies with guinea pigs and marmoset monkeys (Philippens *et al.,* 1998, Philippens *et al.,* 2000a, Philippens *et al.,* 2000b) and protects against lethality and post-intoxication incapacitation after soman. The use of PC in the pretreatment strongly augments the efficacy against soman poisoning. These animals all survived even without post-intoxication therapy. Presumably other factors besides enzyme inhibition play a role in the protection against nerve agents. This idea is strengthened by the finding that there was no correlation between enzyme activity and efficacy of the different treatment groups. Therefore, PC is an interesting compound as an alternative for the current pretreatment PYR against OP-intoxication and should be further investigated.

REFERENCES

Albuquerque EX, Akaike A, Shaw KP, Rickett DL. The interaction of anticholinesterase agents with the acetylcholine receptor-channel complex. Fundam. Appl. Toxicol. 4: S27-S33, 1984.

Albuquerque EX, Aracava Y, Cintra WM, Brossi A, Schonenberger B, Deshpande SS. Structure-activity relationship of reversible cholinesterase inhibitors: activation, channel blockade and stereospecificity of the nicotinic acetylcholine receptor-ion channel complex. Braz. J. Med. Biol. Res. 21: 1173-1196, 1988.

Bakry NM, el-Rashidy AH, Eldefrawi AT, Eldefrawi ME. Direct actions of organophosphate anticholinesterases on nicotinic and muscarinic acetylcholine receptors. J Biochem Toxicol. 3:235-59, 1988.

Dirnhuber P, French MC, Green DM, Leadbeater L, Stratton JA. The protection of primates against soman poisoning by pre-treatment with pyridostigmine. J. Pharm. Pharmacol. 31: 295-299, 1979.

Gordon JJ, Leadbeater L, Maidment MP. The protection of animals against organophosphate poisoning by pretreatment with a carbamate. Toxicol. Appl. Pharmacol. 43: 207-216, 1978.

Johnson CD, Russell RL. A rapid, simple radiometric assay for cholinesterase suitable for multiple determinations. Anal. Biochem. 64: 229-238, 1975.

Kim Y-B, Cheon K-C, Hur G-H, Phi T-S, Choi S-J, Hong D, Kang J-G. Effects of combinational prophylactics composed of physostigmine and procyclidine on soman-induced lethality, seizures and brain injuries. Environ. Toxicol. Pharmacol. 11(1): 15-21, 2002.

Leadbeater L, Inns RH, Rylands JM. Treatment of poisoning by soman. Fundam. Appl. Toxicol. 5: 225-231, 1985.

Myhrer T, Skymoen LR, Aas P. Pharmacological agents, hippocampal EEG, and anticonvulsant effects on soman-induced seizures in rats. Neurotoxicol. 24: 357-367, 2003.

Philippens IHCHM, Melchers BPC, Wolthuis OL. Active avoidance in guinea pigs, effects of physostigmine and scopolamine. Pharmacol. Biochem. Behav. 42: 285-289, 1992.

Philippens IHCHM, Olivier B, Melchers BPC. Effects of physostigmine on the startle in guinea pigs: two mechanisms involved. Pharmacol. Biochem. Behav. 58: 909-913, 1997.

Philippens IHCHM, Busker RW, Wolthuis OL, Olivier B, Bruijnzeel PLB, Melchers BPC. Subchronic physostigmine pretreatment in guinea pigs: effective against soman and without side effects. Pharmacol. Biochem. Behav. 59(4): 1061-1067, 1998.

Philippens IHCHM, Vanwersch RAP, Groen B, Olivier B, Bruijnzeel PLB, Melchers BPC. Subchronic physostigmine pretreatment in marmosets: absence of side effects and effectiveness against soman poisoning with negligible postintoxication incapacitation. Toxicological Sciences 53: 84-91, 2000a.

Philippens IHCHM, Melchers BPC, Olivier B, Bruijnzeel PLB. Scopolamine augments the efficacy of physostigmine against soman poisoning in guinea pigs. Pharmacol. Biochem Behav. 65:175-182, 2000b.

Philippens IHCHM, Groen B, Vanwersch RAP. Direct effects of physostigmine as a pre-treatment in guinea pigs: side effects and efficacy against soman. Proc.TG004 meeting, Medicine Hat, Canada, 2003.

Raveh L, Chapman S, Cohen G, Alkalay D, Gilat E, Rabinovitz I, Weissman BA. The involvement of the NMDA receptor complex in the protective effect of anticholinergic drugs against soman poisoning. NeuroToxicol. 20(4): 551-560, 1999.

Ray R, Clark III OE, Ford KW, Knight KR, Harris LW, Broomfield CA. A novel tertiary pyridostigmine derivative [3-(N,N-Dimethylcarbamyloxy)-1-methyl-Δ3-tetrahydropyridine]: anticholinesterase properties and efficacy against soman. Fundam. Appl. Toxicol. 16: 267-274, 1991.

Sherby SM, Eldefrawi AT, Albuquerque EX, Eldefrawi ME. Comparison of the actions of carbamate anticholinesterases on the nicotinic acetylcholine receptor. Molec. Pharmacol. 27: 343-348, 1984.

Van den Beukel I, van Kleef RGDM, Oortgiesen M. Differential effects of physostigmine and organophosphates on nicotinic receptors in neuronal cells of different species. NeuroToxicol. 19(6): 777-788, 1998.

Wolthuis OL, Groen B, Busker RW, van Helden HHPM. Effects of low doses of cholinesterase inhibitors on behavioral performance of robot-tested marmosets. Pharmacol. Biochem. Behav. 51: 443-456, 1995.

This study is part of the NBC programme, which is sponsored by the Dutch Ministry of Defence.

Chapter 14

BIOMEDICAL SAMPLING FOLLOWING A CHEMICAL WARFARE AGENT TERRORIST EVENT - AN OPCW PERSPECTIVE

Mike Rowell
Organisation for the Prohibition of Chemical Weapons (OPCW)

Abstract: The chemical weapons convention provides for collection and analisis of biomedical samples – blood, urine or othern tissues – in an OPCW investigation of alleged use. The paper provides a synopsis of thye draft report of the OPCW technical working group on biomedical sampling. And how these findings may apply to a atates response to use of chemical agent in a terrorist attack.

Keywords: biomarkers; biomedical sampling; chemical warfare agents; chemical weapons convention; invstigation of alleged use; terrorism

A comprehensive approach to a states response to a chemical terrorism includes having a plan not only for the crisis and consequence management phases of the incident, but also for all elements required for complete resolution of the event. This may include the necessity to definitively establish whether chemical agents were used, to provide supporting evidence to confirm other analyses, or to provide the forensic proof required to support a criminal prosecution. The collection and analysis of biomedical samples – blood, urine or other tissue from affected humans or animals – is one of the means for providing such information. Although current capabilities such as urinary thiodyglycol excretion or plasma cholinesterase activity can be performed, there is scope for far more sensitive and specific assessments that overcome the limitations of these approaches.

1. BIOMEDICAL SAMPLING IN THE OPCW CONTEXT

For the OPCW, the Chemical Weapons Convention provides for the collection and analysis of samples of biomedical origin within the

C. Dishovsky et al. (eds.), Medical Treatment of Intoxications and Decontamination of Chemical Agents in the Area of Terrorist Attack, 123–132.

context of an Investigation of Alleged Use (IAU)[1]. The use of biomedical samples provides a unique and complementary information source in such an investigation, particularly when other sources - primarily environmental sample analysis - are inconclusive. The interaction of chemical agents with biological systems can yield a number of markers of exposure that may be detected at very low levels, including below the level required for a toxicological effect. In certain cases these markers may remain in the body for extended periods of time. In both cases therefore biomedical sampling can be particularly suited to an IAU that occurs weeks or months after the alleged incident, when other detection methods are no longer suitable.

This paper provides a synopsis of the findings of the OPCW Temporary Working Group on Biomedical Sampling which met in the Hague 17-19 November 2004[2]. Although the findings are primarily directed at the technical and practical aspects of providing a biological sampling capability for scheduled chemicals (chemical warfare agents) for the OPCW and its member states, the concepts are broadly applicable to states or other bodies attempting to include biomedical sample analysis capabilities within their overall chemical terrorism response capabilities.

2. BIOMEDICAL SAMPLING IN RESPONSE TO AN EVENT

In the emergency response phase of an incident, initial determination of the agent often occurs on site, particularly if there is a significant release of one of the classical threat agents. The use of any of the well established portable technologies such as wet chemistry, IMS or portable GCMS each provides an initial guideline on the class or type of agent used. If humans (or in some cases animals) are exposed then the clinical findings and response to treatment can also provide further information that can be added to the overall picture. Where environmental samples can be taken, off site analysis using more sophisticated laboratory equipment such as research grade GCMS/LCMS can provide a more certain confirmation of these findings, or, through their greater range and lower limit of detection, an initial determination.

For less well defined incidents however, these detection systems may be inadequate. Portable chemical detectors may not be able to be deployed to the site, not detect the agen, or give inconclusive results. Clinical findings may be non-specific, present in an atypical manner, or for example in the case of sulphur mustard, have a latency period that delays firm pattern recognition. Due to the physico-chemical properties of the agent or the time between release and collection, environmental samples may have low agent levels or sufficiently high contaminants to prevent adequate results.

---

[1] Verification Annex, Part XI, paragraphs 16, 17

[2] Draft Report – Temporary working Group on Biomedical Samples

If these methods do prove inconclusive, biomedical sample analysis may provide a unique method for establishing exposure. Due to both the complex technical requirements and the strict forensic approach used it is unlikely that this information will be available in the early phases of response to such an incident. Biomedical sample analysis does however offer another method by which proof of the use of chemical agents can be provided, and thus has a potentially significant role in the overall preparation of a capability for the response to a terrorist chemical agent attack.

3. SAMPLE COLLECTION

Biomedical sample analysis relies on appropriate sample collection. Although any result is unlikely to guide the response phase of the incident, the emergency response should include consideration of the collection of these samples. The best results are expected to be generated from samples taken as early as possible, and from patients considered to have had the greatest exposure. There is however little clear data available on the urgency of analysis in relation to the stability of markers, and on the effects of lag-times for bringing samples to the laboratory.

The OPCW has developed a framework for the collection and transportation of biomedical samples[3]. It is believed that this document provides a solid basis for further development in this area, and these procedures will be tested during an upcoming major exercise in 2005. Some of the key issues addressed include the requirement for samples to be collected by medical staff only, the contents of the sampling kit, the packaging requirements and conditions required for transport (such as refrigeration), and the chain of custody mechanism. The reception and sample preparation is not included in this document but is the logical next phase for review, and will rely heavily on the capability (or desire) of designated laboratories to deal with biological materials.

An area of sample analysis that is likely to provide some practical difficulty regards informed consent, and the use of medical information gained from the samples. The person providing the sample should be made aware of the risks and benefits of the procedure, what will happen to the sample, who will have access to the analysis and accompanying information, and for how long, and what will be done with the results. Importantly, from the medical ethical view, the patient should also know as soon as practicable what the results of analysis of samples were, and what they mean. Most countries have legislation in place that regulates access to and use of medical data. These requirements will need to be considered against the very real likelihood that they may conflict with the security or criminal/judicial implications of releasing this information. Whilst clearly applying to a state responding to a terrorist incident, it is

---

[3] Draft Working Instruction for the collection of biomedical samples during an investigation of alleged use.

also a major hurdle for the OPCW to overcome within the sensitive political environment within which the OPCW may be conducting an IAU.

As seen in table 1, "Sampling and Analysis of Biomedical Samples for the Presence of Chemical Agents: Summary of Key Points", most research has been carried out using either venous blood or urine as the sample type. Although other tissues could be used, blood and urine are used for practical reasons as much as technical. These samples are common, easy to obtain from both humans and laboratory animals, and therefore have most of the validated research based on them. During or after an incident they may be legitimately taken for other pathology tests related to treatment as well as agent identification Blood and urine analysis are also in themselves complementary, and have different attributes that may better suit different scenarios or laboratories. In general, urine analysis offers a simpler approach, is less expensive and is suitable for detection in the earlier period after exposure. Blood analysis that focuses on protein adducts for example, is likely to be more sensitive, and to yield results from samples taken much longer after the event.

4. BIOMEDICAL SAMPLE ANALYSIS COMPARED TO ENVIRONMENTAL SAMPLE ANALYSIS

The general requirements for biomedical sample analysis are in many regards similar to those for environmental sample analysis. Procedures should be well defined, cost-effective and well validated. The results obtained should be accurate and able to be combined with clear guidelines on interpretation of results. The laboratories which perform the tests should have well documented Quality Assurance/Quality Control procedures and undertake regular exercises to maintain accreditation for their work. The extent to which these goals is met will determine whether the sampling procedure will fulfil the primary objective - of providing solid evidence regarding use (or presence) of chemical agents.

The OPCW expert group found however that the procedures for environmental sample analysis were too inflexible to achieve the desired outcome. A separate approach would be needed to enable a system that would provide the necessary level and quality of information, but be practical and also recognise the very infrequent likelihood of a requirement for this capability. Compared to the current OPCW approach to environmental samples, that for biomedical samples would likely use a laboratory with a research rather than a “routine” focus, the accreditation procedures would be less stringent, and the methods used would focus on high quality results at the lowest limits of detection, rather than high throughput. It is not expected that more than a few laboratories will wish to maintain all or some of the necessary expertise to perform these tasks.

As the OPCW biomedical sampling requirement is of the lowest level of detection possible, the samples would be presented to the lab with as much relevant information as possible, to enable the most appropriate test or methodology to be selected. This includes times of suspected release and sample collection, clinical affects seen, medical characteristics of the donor (age, sex etc), whether treatment had been given and the like. This is in contrats to environmental samples, which are often presented with little additional information in order to maximise the integrity of the analysis and interpretation.

5. TECHNICAL ASPECTS OF BIOMEDICAL SAMPLING

The technical elements of biomedical analysis are not the focus of this paper however a brief mention of the areas identified by the OPCW expert group is warranted. A summary of the conclusions as they relate to these areas is presented in table 1.

A major area for research includes further establishing key biomarkers of exposure. Because each agent may have multiple metabolites, degradation products, protein or DNA adducts, there are many potential biomarkers. Establishing which are most suitable depends on multiple factors including cost, the analytical standards required, and laboratory expertise. The methodology for assessing each biomarker would then have to be validated and a QC system put in place.

Another area to be resolved would be to determine the interpretation criteria for each method. Unlike environmental analysis, which aims for a clear, high quality full spectrum scan confirmed by a complementary technique, the analysisofmetabolitesoradducts may not have a readily available or as reliable confirmatory technique. The implementation of a process that rates biological analyses, such as that used by the EC for identification of chemicals in animal products, was suggested by the expert group as a possible means to establish how well an analysis has been able to be confirmed[4].

6. BIOMEDICAL SAMPLING IN THE OPCW CONTEXT

The OPCW is engaged in development of a biomedical sampling capability primarily in support of its role in Investigations of Alleged Use. It has prepared guidelines on the collection of samples in the field. As an interim measure for the analysis component, proposals have been formulated that would identify laboratories of Member States that are working in this field, and allow them to receive, analyse and store samples collected during an investigation. This is a departure from normal OPCW practices, but is considered the most practical measure were a biomedical sampling capability required in the near future.

A possible future role may be in maintaining a forum for those laboratories with an interest in this field, facilitating the development of

---

[4] European Commission decision of 12 August 2002 implementing Council Directive 96/23/EC concerning the performance of analytical methods and the interpretation of results.

methods and other technical elements, and compiling or maintaining a database of these techniques and capabilities for member states.

7. CONCLUSION

Although not yet a mature science, the development of a practical biomedical sampling capability is underway. At present, the OPCW does not have such a capability, and the capability in Member States in this field is also still very limited. Despite a number of practical and technical challenges, biomedical sample analysis has the potential however to be an extremely sensitive and specific method of establishing credible forensic information on the alleged use of chemical agents.

Table 1. Sampling and Analysis of Biomedical Samples for the Presence of Chemical Agents: Summary of Key Points

NERVE AGENTS

Sample type	Key biomarkers	Recommended analytical methods	Standards	Advantages	Disadvantages
Blood	Cholinesterase activity			Rapid, available in the field	Does not identify the OP; False positive results; Only relatively high levels of activity depression are detectable
Blood	Fluoride reactivation method: * Phosphorylated BuChE (+ other proteins)	GC-MS GC-HR-MS With large volume injection;	Phosphonofluoridates i.s.: deuterated OP or plasma exposed to deuterated OP	Easily accessible internal standards and reference compounds; LOD 10 pg/ml (0.05-0.1% BuChE inhibition)	Not applicable to all OP's

NERVE AGENTS *(continuation)*

Blood	Analysis of phospho-rylated peptides: Phosphyla-ted BuChE	LC-MS-MS (after enzymatic digestion of modified cholineste-rase.	Phosphylated nonapeptides; i.s. plasma exposed to CD3-OP	Covers all OP's LOD: 1-5% BuChE inhibition	Expensive instrumentation and reference compounds
Urine / serum	Hydrolysis products: Alkyl methylphos pho-nic acids (does not include Tabun)	GC-MS-MS LC-MS-MS	(derivatized) alkyl methylphos-phonic acids; i.s.: CD3 analogues	High levels shortly after exposure. Easily accessible internal standards and reference compounds; LOD 0.2 – 1 ng/ml	Relatively short window of detection

(* SOP available)

SULPHUR MUSTARD

Sample type	Key biomarkers	Recommended analytical methods	Standards	Advantages	Disadvantages
Urine	TDG TDGO β-lyase metabolites	GC-MS-MS LC-MS-MS	(deuterated) TDG TDGO β-lyase metabolites	Relatively easy synthesis of analytical standards LOD: 1 – 0.1 ng.ml	Short window for detection (max. 2 wks) TDG and TDGO present in unexposed persons.
Blood	<u>Protein adducts:</u>	Chemical or enzymatic digestion, followed by:		Longer window for detection.	More demanding analytical methods.
	N-terminal valine on Hb*	GC-MS or GC-MS-MS	PFP-thiohydantoin of valine adduct i.s.: d8-sulfur mustard-globin	Can be run on a benchtop GC-MS LOD: 100 nM human blood, in vitro	Laborious
	histidine residues on Hb.	LC-tandem MS	His-adducts I.s.: d8-sulfur mustard globin	His adduct also present in proteins of other tissues	Laborious
	Cysteine residue on albumin*	LC-tandem MS	Cys*-Pro-Phe i.s.: d8-sulfur mustard plasma	Easy work-up of sample LOD: 1 nM human blood, in vitro	Alkylated tripeptide required as analytical standard
	Asp. Acid/glutamic acid residues on blood proteins and keratin	GC-MS	Thiodiglycol (derivatized); i.s.: d8-thiodiglycol	Easy work-up of sample LOD: 25nM	
	DNA adducts:				
Urine Blood	Alkylation of deoxyguanosine (N7)	LC-MS-MS for N7-HETE-guanine	N7-HETE-guanine + d8 derivative as i.s.	Macromolecule present in all bodily tissues.	Shorter window for detection than protein adducts
	Alkylation of deoxyguanosine (N7)*	ELISA assay for N7-HETE-guanosine-5'-phosphate	Sulphur mustard exposed DNA	Low cost LOD 50 nM human blood in vitro	Monoclonal antibody required. Less specific than MS-based methods

*SOP available

LEWISITE

Sample type	Key biomarkers	Recommended analytical methods	Standards	Advantages	Disadvantages
Urine	CVAA	Solid phase microextraction headspace sampling, followed by GC/MS with EI ionisation	Deuterated CVAA Phenylarsine oxide	LOD: 500 ppt	Short window for detection. Lack of validation in human samples.
Blood	CVAA (globin bound and free)	GC-MS	As above	LOD: 1nM	As above

PHOSGENE

Sample type	Key biomarkers	Recommended analytical methods	Standards	Advantages	Disadvantages
Blood	<u>Protein adduct:</u> Albumin peptide	LC-MS-MS	Whole blood treated with known phosgene concentrations	Specific and sensitive LOD: 1uM	Standards not easily available

CYANIDE

Sample type	Key biomarkers	Recommended analytical methods	Standards	Advantages	Disadvantages
Blood	CN itself	GC	HCN	LOD: 100ppb	Background levels (smokers, pollution etc)
Urine	Cystine adduct SCN 2-amino-thiazoline, 4-carboxylic acid	HPLC GC–LC	SCN	(SCN) LOD: 3ug/ml 0.3 uM	As above

BZ

Sample type	Key biomarkers	Recommended analytical methods	Standards	Advantages	Disadvantages
Urine	BZ, BA Q	LC-MS-MS	BZ, BA, Q	Rapid and sensitive LOD: 1ppb	Standards expensive and not easily obtainable

Chapter 15

PARTICULARITIES IN RESEARCH, PRODUCTION AND ACQUISITION OF THE PHARMACEUTICAL PRODUCTS FOR NBC MEDICAL PROTECTION

Florin Paul

Romanian Ministry of National Defense, Romania

Abstract: Medical research is a branch of biological research, as a fundamental area of scientific research. The results of medical research could be pharmaceutical products, diagnosis methods, surgical or investigational techniques, that are used for medical protection in an NBC environment. The main fields covered by biomedical research in NBC protection are: toxicology, radiobiology, microbiology, pharmacology, pharmaceutical techniques, internal medicine etc. New drugs and vaccines, diagnosis, treatment and prophylaxis methods are the final results of these activities. Rarely, depending on the level of technological development, there is research in the field equipment for pathogens or noxious chemicals identification. The steps required to obtain a new pharmaceutical products are many and often very difficult to go through. The first step is biochemical synthesis of the experimental product, followed by preclinical experiments. These include "in vitro" tests and experiments on laboratory animals that could be forbidden by law, especially if they kill the animal. A very sensitive step is clinical experiment. This supposes to follow all legal aspects regarding experiments on human beings. Also teratogenic, oncogenic, citotoxic effects must be identified, in order to avoid harmful effects on the treated persons..The minimum time to introduce a new drug or vaccine in the medical practice is about ten years. In the biomedical research for NBC pharmaceutical products a very important issue is to find the producer and the funds for them. Due to such difficulties these products are named "orphan products". This very important aspects requires the involvement of government and public authorities, because the lack of "orphan products" is followed by losses of human lives.

Keywords: Orphan drugs; registration and usage of pharmaceutical products; vaccines and antibiotics in biological prevention

C. Dishovsky et al. (eds.), Medical Treatment of Intoxications and Decontamination of Chemical Agents in the Area of Terrorist Attack, 133–141.

1. INTRODUCTION

Recent worldwide terrorist acts and hoaxes have heightened awareness that incidents involving weapons of mass destruction (WMD) may occur everywhere. This fact requires the development of preparedness programs to train and equip emergency services and emergency department personnel in the management of large numbers of casualties exposed to nuclear, biological, or chemical (NBC) agents. Hospital pharmacies and national pharmaceutical stockpiles will be required to provide antidotes, antibiotics, antitoxins, and other pharmaceuticals, in large amount and have the capability for prompt procurement. There is no doubt that both physicians and pharmacists should became knowledgeable in drug therapy of NBC threats with respect to nerve agents, cyanides, pulmonary irritants, radionucleotides, biological agents as anthrax or botulism, and other possible WMD.

Protection against nuclear, bacteriological and chemical hazards and emergency treatment of induced toxic effects is based on antidotes and special pharmaceutical products, other than current drugs. These pharmaceutical products should be included in the "orphan drugs" group that includes also the medicines used in more than 3000 rare and very rare diseases. Antidotes and other pharmaceutical products as vaccines or antitoxins are called "orphan drugs", in the sense that, following their sporadic use in normal times, their production is not profitable for pharmaceutical companies.

The development of drugs for these diseases, intended for a limited number of patients, often requires considerable research, and is consequently costly. A particular approach of ethical, political and economical problems relevant for development and disposal of orphan drugs is also required.

The registration of this special means encounters substantial difficulties in almost all countries, following their inclusion in the category of usual drugs. A lot of requirements concerning a very wide, preclinical and clinical investigation, which practically hinders registration of antidotes and makes no sense in case of these special pharmaceutical products.

Complying with such requirements can lead to drastic reduction of the availability and to long delays in obtaining the approval for registration and industrial production.

Mass casualty in NBC disasters, require immediate availability of antidotes and special pharmaceutical products that can save the live of affected population.

In respect to this issue special legal provisions were arranged in the USA and European Union to facilitate the production and use of orphan drugs.

These stipulations would provide to society a tool for imposing on pharmaceutical companies to support an important part of these expenses for production of "less profitable" orphan drugs, especially those for NBC medical protection of the population.

2. DISCUSSION

2.2. Particularities of NBC orphan drug research and production

The lack of financial profit of the production of orphan drug production is the main obstacle in achievement of the protection task. NBC orphan drugs (NBC-OD) are necessary in large amount, only in "critical situations" (war, natural or technological disaster, terrorist attack etc.). Practically all NBC-OD, with very few exceptions, like anthrax vaccine, have no "civil" use, as drugs.

Another serious difficulty arises from the assimilation of NBC-OD with common drugs.

Almost all NBC-OD do not fulfill the requirements for registration as drugs; their registration must be led by other rules.

The technical difficulties in research, production and use of NBC-OD are less known. These difficulties come first from special requirements imposed to NBC-OD.

The requirements for treatment of chemical and antitoxin NBC-OD are:

- increased stability;
- self-administration;
- universal action;
- high efficiency;
- instantaneous onset of action.

Preventive antidotes, protectors and decorporators must have:

- lack of incapacitating effects;
- oral or percutaneous administration;
- lack of adverse effects after long term and many administrations;
- longer biological half-life after administration.

The requirements for vaccines, immune treatment prophylaxis products and antibiotic are:

- rapid efficacy and high immunogenesis effects;
- availability and long shelve time;
- stability of microbial strain used in vaccine;
- covering the entire antigenic profile of the targeted pathogen;

- wide spectrum for antibiotics and no special condition for preservation;
- appropriate conditions for storage, transportation and usage for large areas;
- quality control for immune serum in order to avoid transmission of other diseases, as hepatitis B and C, HIV infection etc.

All these requirements are difficult to be achieved, some of them being rather contradictory.

2.3. Particularities in acquisition of antidotes – registration, delivery and usage

Antidotes are listed and classified in accordance with their effectiveness and availability (Table 1, 2, 3). Antidotes reduce the overall burden of health service in managing of poisoning cases.

In developing countries that lack adequate facilities for intensive therapy of poisoned people, antidotes may be more essential in the prevention and treatment of poisoning. But availability of antidotes is different from one country to another. In developing countries, physicians reported difficulties in obtaining even common antidotes. Even in industrialized countries, could be noticed administrative difficulties and the lack of suitable drugs (pharmaceutical formulation, concentration etc.).

A very important issue is that pharmaceutical companies may hinder the access to certain antidotes for various reasons.

Difficulties in obtaining of adequate availability arose from three interrelated areas: scientific and economical, regulatory and administrative requirements, and managing distribution in crisis.

Governments and chemical industries are responsible for ensuring comprehensive scientific studies for regulatory authorities to accept registration of effective antidotes.

In the same time pharmaceutical companies involved in production of antidotes must be encouraged to register their products in their own countries. In this respect, it is very helpful that administrative procedures of registration and disposal of an antidote, for example, will comply with international rules regarding the orphan drugs.

Pharmaceutical companies will manufacture and supply antidotes only if they are encouraged by adequate economic refunds for their investment and by simple registration procedures.

Particular aspect regards common drugs (active substance) used in therapy that could be used successfully as antidotes, but in different formula or concentration. In that condition additional authorization is required. Procedure for authorization should be simplified. Authorities need to accept similar criteria for registration of a new antidote (less comprehensive that than for normal drug) as, for example, for anticancer or anti AIDS agents because of the special conditions of their use.

Table 1. PHARMACOLOGIC PROFILE FOR NBC – ORPHAN DRUGS. (Efficacy and availability)

- High and specific efficacy
- Rapid effect
- Long-lasting action (hours)
- Suitability (easy to administer)
- Chemical stability (stored unaltered for long time)
- Availability in crisis in the required quantities
- Provided on pre-planned basis to individuals, to intervention team etc

Table 2. NBC – ORPHAN DRUGS, ARE THEY AVAILABLE?

- No (or very low) prescription in normal circumstances ("orphan drugs")
- Manufactured in small amounts (or not manufactured)
- Valid for no more than 1-3 years
- Quantifying drug requirements not possible, except high risk areas or predictable circumstances (terrorism, for example)

Table 3. ANTIDOTES CLASSIFICATION VS. AVAILABILITY AND EFFECTIVENES.

Gr. I. (WHO, IPCS, EC) Antidotes useful in treatment of poisoning

- Availability:
  - **A.** Required to be immediately available (within 30 minutes)
  - **B.** Required to be available within 2 hours
  - **C.** Required to be available within 6 hours
- Effectiveness
  - **1.** Effectiveness well documented
  - **2.** Widely used, but requiring further research concerning effectiveness and/or indications

2.4. Particularities of acquisition of vaccines

The development of vaccines against rare emergent infectious diseases is hampered by many disincentives.

Vaccine development involves a substantial investment in time, effort, and resources. Any public or private research and producing facility should allocate huge financial and human rescues when development of vaccine is decided. The cost from research to licensure, the risks inherent in vaccine development (e.g. technological constraints, regulatory approval) and short- or long-term evaluations of scientific and financial results may constrain this activity. In the developing world, the price has been a major impediment to the introduction of new vaccines.

Reliable information on the epidemiology, disease severity, and effect on public health is essential to sustain the need for a vaccine. The authorities must develop the policy to prevent infectious diseases and in the same time countermeasures against effects of biological weapons attack.

Development of orphan vaccines is guided by the limited need for or markets potential of the product, with the accompanying regulations, as well as the specific characteristics of the vaccine and those who need it. After September 11th, 2001, the threats of biological attack open perspectives for acquisition of new vaccines and immune-prophylaxis products.

However, research for new effective products needs a long time, and the development of any orphan vaccine should be broadly supported by measures to increase the awareness of immunization benefits at three levels – the decision-makers, the care-givers, and the patients.

Developments in biotechnology have created the promise of prevention for many more infectious diseases and chronic diseases and build the confidence in acquiring new effective measures against biological warfare agents.

2.5. Particularities in stockpiling and delivering of pharmaceutical products

The availability of an NBC-OD is highly dependent on its manufacturing, delivery procedures, and economic power of society.

The costs of procurement of NBC-OD is a sensitive issue, looking to developing countries that can not afford high expenditures even in crisis situation.

On the other hand it is practically impossible that all countries will develop production facilities for the whole range of NBC-OD.

In circumstances of increasing the threat of international terrorism, and attack with WMD, the regional and international cooperation becomes mandatory. A part of this cooperation is the availability of NBC-OD for affected population. In some cases, like biological attack, the affected area could be larger, pathogens crossing the political or administrative borders.

If certain NBC-OD are not available from local manufactures and must be imported there are two alternative solutions: to establish a manufacturing facility (or a pharmacy laboratory) supported with government funds or the establishment of a central agency for import and distribution of antidotes, under governmental control. The decision depends on the economical and technological capabilities.

Storage facilities for NBC-OD require specific conditions:

- distance from medical facilities and transportation facilities (airport, roads);
- inside temperature and humidity;
- communications;
- building safety in case of WMD attack, natural or technological disaster;
- capacity of storage;
- real time of intervention.

The amount and the type of NBC-OD reserve depend on:

- size and geographical profile of the exposed area to WMD attack;
- nature of potential NBC hazard;
- number and density of population in the affected area;
- distances to medical care facilities from theater, communications etc.

3. CONCLUSIONS

- international and regional consensus in fighting against WMD threats and elaboration of a common strategy of intervention in crisis situation;
- governmental support for developing the facilities for production, import and storage of NBC-OD;
- simplifying the methodology of registration and approval for NBC-OD, and special legal provisions in this respect is mandatory;
- development of international and regional programs for scientific research, production and distribution of NBC-OD appear as an urgent requirement in fighting the NBC threats and international terrorism.

REFERENCES

1. Burda AM, Sigg T. Pharmacy preparedness for incidents involving weapons of mass destruction. Am J Health Syst Pharm 2001, Dec, 1, 58(23): 2274-84;
2. Food and Drug Administration, USA. Improving Public Health through Human Drugs. Report 2001. http:/www.fda.gov/cder/cder.org.htm;
3. Lang J, Wood SC; Development of Orphan Vaccines. An Industry Perspective. Emerging Infectious Diseases, 1999, 5(6): 749-755.
4. Matherlee K. The Public Stake in Biomedical Research: a Policy Perspective. National Health Policy Forum, November 1999, USA. http:/www.cdc.gov/nih;
5. Mircioiu C., Voicu V. Self protection of pharmaceutical companies against terrorist attacks as core of large area population protection. Proceedings of CB Medical Treatment Symposium Industry I, 151-153, 1998, Dubrovnik, Croatia;
6. Schwartz B., Rabinovich NR; Stimulating the Development of Orphan (and Other) Vaccines. Emerging Infectious Diseases, 1999, 5(6):832;
7. Shah ND; Vermeulen LC; Santell JP and all. Projecting future drug expenditures. Am J Health Syst Pharm, 2002, 59(2):131-140;

8. Voicu V., Mircioiu C. Antidotes – Individual protection means: drugs and special means. Proceedings of CB Medical Treatment Symposium Industry I, 315-326, 1998, Dubrovnik, Croatia;
9. Watson R. Health ministers to help orphan drugs. British Medical Journal, 1995; 310:1557-1558;
10. Wong SH; Challenges of toxicology for the millennium. Ther Drug Monit, 2000, 22(1):52-57.

Chapter 16

TOXICOKINETICS IN HELPING OF DIAGNOSES AND TREATMENT OF CHEMICAL POISONING

Manana Juruli
Georgian Environmental and Biological Monitoring Association, Tbilisi, Georgia

Keywords: biological monitoring; biological threshold limit value; body fluids; chlororganic; elimination; heavy metals; phosphororganic

1. INTRODUCTION

The increasing prevalence of toxic substances in the environment has become a matter of public interest as a complex public health problem. The main goal of preventive toxicology is to reduce exposure to potentially harmful substances to the levels that do not cause toxic effects. First step on this way is the assessment of exposure. Traditionally, this has been done through air sampling and measurements. But it is not quite enough to evaluate the real picture of the exposure, as concentrations of a chemical in the ambient air are not always very closely related to the amounts absorbed. Concentrations in the air are seldom stable but fluctuate with time and are different in various locations. Many chemicals are effectively absorbed through the skin, and this route of exposure is not at all related to concentrations of the chemical in the air. The particle sizes of various dusts profoundly affect their deposition patterns in the airways and thus their absorption. Personal life style and habits vary, and individuals may absorb different amounts of chemicals in superficially similar conditions It is thus quite evident why it is necessary to use biological monitoring of harmful substances to evaluate exposure and to take follow- up preventive measures to reduce the health impact.

In order to fill in these blanks and to give a more real reflection of the picture of exposure, the biological monitoring of harmful substances should be carried out and, especially its kinetic properties in the body should be established. Toxicokinetic models allow to estimate the Biological Threshold Limit Values (Reference Values), which serve to protect the individuals effectively from chemical–induced health risks.

We have studied the toxicokinetics of heavy metals (arsenic, lead, manganese and copper), as well as of organophosphorus and clorinated pesticides. It has been shown that they are all characterized by different

C. Dishovsky et al. (eds.), Medical Treatment of Intoxications and Decontamination of Chemical Agents in the Area of Terrorist Attack, 143–144.

toxicokinetics parameters. Heavy metals and chlororganic pesticides have long half-times and they have the tendency to accumulate in the body. Phospo-organic compounds have much more shorter half times but they are easily absorbed through the skin.

The aim of this presentation is to show arsenic toxicokinetics after various single doses in rats.

2. METHODS

Oral single doses (3, 30 and 100 mg/kg) of arsenic trioxide were given to rats. The amount of As in blood and urine was determined at various times during the experiment until it had completely disappeared from blood and urine.

3. RESULTS

The arsenic concentration in blood were eliminated according to a three-phasic pattern. Arsenic excretion in urine does not seem to be influenced by the varying doses of administered arsenic trioxide. It is shown that 67 % of administrated arsenic trioxide (100 mg/kg) is excreted by kidneys with $T_{0.5}$ = 3.6 days; 74 % is eliminated by urine with $T_{0.5}$ = 2.4 days (30 mg/kg); and 68 % is eliminated by urine with $T_{0.5}$ = 0.8 days (3 mg/kg);

4. CONCLUSION

It has been shown that all doses of arsenic trioxide are characterized by different toxicokinetics parameters. Arsenic compounds have long half-times and the tendency to accumulate in the body. The excretion rate decreased with decreasing blood concentration. The present study confirms the ability of toxicokinetic models to improve the study of various toxic substances and to estimate the Biological Threshold Limit Values.

Part 3

DEVELOPMENT OF PERSONAL DECONTAMINATION IN CASE OF INTOXICATION WITH CHEMICAL AGENT

Chapter 17

ASPECTS OF DECONTAMINATION IN CASE OF RELEASE OF TOXIC SUBSTANCES OR USE OF CHEMICAL WARFARE AGENTS

Tsvetan Popov[1], George Popov[2]
[1]State Agency for Civil Protection, Republic of Bulgaria
[2]Kingston Environmental Laboratory, USA

Abstract: Some explanations, definitions and guidelines for response and consequence management of situations with hazardous substances are given. Preparedness, decontamination activities and measures undertaken in further development of the situation in contaminated areas are underlined.

Keywords: decontamination, individual protection equipment, decontaminants, antidotes, decontamination kits

1. DISCUSSION

The increase of terrorist activities during the last decades implies the need for implementation of new, strengthened measures for safety and protection of the population and critical infrastructure. Threat, release or use of toxic chemical substances and biological agents is real and represents an existing danger which calls for preparedness and organization of the authorities, units and means for response and consequence management in emergency situations.

The scope of this article is to emphasize the preparation, procedures and decision-making process in case of release of toxic substances or use of chemical warfare agents.

In case of accident or emergency situation involving hazardous materials the preliminary information on the situation is very important for a rapid and adequate response. If the situation concerns industrial or transport accident with known substances release information could be sufficient to prepare proper:

- Individual protection means
- Decontamination equipment and decontamination substances

C. Dishovsky et al. (eds.), Medical Treatment of Intoxications and Decontamination of Chemical Agents in the Area of Terrorist Attack, 147–152.

- Antidotes, medicines, etc.

In most of the cases preliminary information is insufficient and both rescue units and consequence management authorities have to be prepared for a scenario with high potential risk.

The first assessment on the situation includes:

- Primary information – visible effects and events after release or use of toxic substances
- Meteorological conditions – temperature, wind speed and direction, vertical stability of air, relative humidity, etc
- Risk assessment using computer software (ALOHA, ARIPAR, etc), database of toxic substances (NIOSH, ERICards) or other methods for mitigation of spill or release (Emergency Response Guidebook 2000)

Logistics should be provided with proper equipment and means for deployment:

- Individual protection means and decontamination products in reserve – replacement of both isolating and filtrating protective clothing, canister filters and SCBA cylinders
- Communication equipment
- Need for electrical and water supply (attention to be kept when combustible or explosive materials are involved)
- Rope or visible band, signs for hazardous substances
- Garbage plastic bags
- Steel drums for the bags
- Plastic or other synthetic covers and layers for spills and decontamination zone
- Vacuum cleaners (for water-reactive substances)
- Other practical means for response
- Drinking water
- Helmets
- Projectors

Response to the situation requires according to the preliminary information and first assessment on the situation:

- Evacuation of endangered population
- Rope off and secure the contaminated area

The First responders provide detection in contaminated area with detection team established on at least two persons equipped with:

- Breathing apparatus (BA) and full chemical protection suits
- Detection and identification means and instruments
- Communication equipment
- Sampling kits

Before entering the contaminated area the detection team should be instructed according to the actual information on the situation.

The detection team gives information about the toxic substances or groups of agents detected and prepares first report on the situation:

conditions, injured people found. During the detection process sampling of air, water, soil, spills, etc is performed for further analysis in field or stationery laboratory.

Figure 1. Detection and decontamination equipment. Bulgarian crisis intervention team. European Disaster Response Exercise, Austria, October 2004.

Information provided to the rescue units is required to prepare proper equipment – individual protection means needed (NIOSH Certified Equipment List database or manufacturer's instructions should be observed), antidotes and pretreatment, stretchers, individual protection means for injured and casualties, personal decontamination kits, etc.

The rescue units should receive instructions according to the information given by the detection team before entering the contaminated area.

First responders engaged with rescue operations provide first aid to injured persons into the contaminated area, decontamination with universal decontamination kits and put individual protection means to avoid further contamination of respiratory tract and body of the injuries.

When exact information on the chemical agent, concentrations and level of toxicity is available reassessment of the situation and further development of the operations is essential for:

- Mitigation of consequences with computerized software or other methods
- Rearrange the perimeter of guarded area
- Providing of proper decontamination procedures

Before starting activities in contaminated area and decontamination procedures some rules should be observed:

- No one enters the contaminated area without proper individual protection means
- No one leaves the contaminated area before the decontamination is provided
- Decontamination substances and ready-to-use solutions as well as decontamination means should be deployed before completing the detection and identification of the unknown substances
- Medical examination for all which come out of the contaminated area after decontamination took place

While organizing the decontamination and choosing the location some conditions should be considered:

- Weather conditions
- Wind direction
- Slope of the ground
- Availability of water
- Availability of power and lighting
- Location of drains and sewers
- Marked entry and exit points

Before exiting the contaminated area on-site decontamination should be performed to the injured persons and self-decontamination of rescue units.

The decontamination procedures include:

- Decontamination of external surface of samples' holders
- Decontamination of rescue units
- Decontamination of the equipment and vehicles situated in contaminated area
- Field decontamination

Decontamination equipment could be various but mostly is based on:

- Spraying decontamination kits and pumps
- Shower systems and tents
- Decontamination complexes (vehicle based)
- Vehicles for screening with decontamination solutions

Figure 2. Decontamination system. Bulgarian crisis intervention team. European Disaster Response Exercise, Austria, October 2004

Decontamination of the rescuers is accomplished according to the type of individual protection means and commonly includes procedures:

- Performed from the upper part of the body to the boots (canister filter and the facepiece of the gas mask at last)
- The removal of the clothing is followed by stepping ahead

Requirements to decontamination substances and solutions for decontamination of vehicles, roads and fields are efficiency and prize, due to the large amount of substances to be used for field decontamination.

Field decontamination depends on the amount of released substance and the level of toxicity and could be accomplished naturally (due to the processes of evaporation, hydrolysis and others) or to be performed using decontamination systems as AGV-3U or Karcher:

- Water solutions of hypochlorites
- Water solutions of NaOH, Na2S, ammonia, etc

In some cases groundsurface removal is required (heavy metals, cyanides, arsenic compounds) for further remediation and treatment.

In spite of the concept that "The human comes before the nature" some precautious measures and environmental considerations could be undertaken:

- Cover the surface where decontamination and rinse is provided
- Using of small pools

- Reservoirs of collecting used decontaminations solutions and rinse water

For further treatment the equipment and IPE should be:

- Packed and sealed (double and triple if it's needed)
- Labeled with appropriate information (contaminant, used decontamination solutions)

Preparedness and organization for response in critical situations involving toxic chemical substances requires a limited amount of decontamination powders, solid substances or solutions to be stockpiled and to be applicable.

According to speed limitations, decontamination should stress on thoroughness, not on speed.

Speed is important when:

- Victim is involved (even then all precaution measures should be undertaken)
- Contaminated civilians or other emergency response workers are involved
- Protective suit of rescuer is split or damaged and a rescuer is injured
- BA failed to supply fresh air for breathing

2. CONCLUSIONS

1. Acknowledgement of the chemical substances and their toxicity is needed for adequate response.
2. Rescue units should be trained to provide first aid to injured persons in contaminated area.
3. Training, exercises and practices are needed to improve the capabilities for the rescue teams to respond to critical situations and accidents.
4. Persistent research and development of the new decontamination substances, equipment and means is important to facilitate the decontamination procedures and to improve the efficiency of the decontamination.

REFERENCES

1. "Guidelines for Decontamination of Fire Fighters and Their Equipment Following Hazardous Materials Incidents", NFPA International 1997.
2. Adamovich I.S., N.I. Ivanov, "Basis of Deactivation, Decontamination and Disinfection", Moscow 1968.
3. Kaufman J.W., Hastings Sh., "Respiratory Demand During Rigorous Physical Work in a Chemical Protective Ensemble", Journal of Occupational and Environmental Hygiene, 2005.

Chapter 18

PERSONAL DECONTAMINATION IN CASES OF CHEMICAL TERRORIST ATTACKS

Jiří Matoušek
Masaryk University, Faculty of Sci., EU Res. Ctre of Excellence for Environmental Chemistry and Ecotoxicology, Kamenic126/3 CZ-625 00 Brno, Czech Republic

Abstract: Main features of the R&D resulting in the new means for primary decontamination of chemical warfare agents based on the chemisorption principle introduced into the Czech Army´s individual decontamination mean IPB-80 and into the Czech Civil Protection first aid kit ZPJ-80, and in the upgrading of sets for secondary decontamination PCHB-60-P and PCHP-60-P are presented. First results of R&D on universal solutions for detoxification of super-toxic lethal chemical warfare agents on human skin are shown and discussed.

Keywords: chemical warfare agents, decontamination, detoxification, human skin, chemisorption, oxidation, detergents, alcoholates, aprotic solvents

1. INTRODUCTION

To the most important rescue measures after contamination of body surface with supertoxic lethal chemical warfare agents belongs decontamination (detoxification) on the lowest tactical level, i.e. at the individual level. Skin contamination with agents possessing extreme percutaneous toxicity and high penetration rate, such as VX, V-gaz, GF, and like stresses the urgency and effectiveness of *primary decontamination* having thus life-saving importance.

As the **first Czechoslovak post-WW-II generation** of means for personal decontamination at the lowest tactical level, i.e., at the individual first-aid level, we considered the two-solution system produced in Czechoslovakia according to the Soviet-originated model IPP-51 . It was introduced under the acronym **IPB-60** into the Czechoslovak Army, and under more simple modification in the first aid kit **OZB** into the Czechoslovak Civil Defence in the early 1960s. The same system was used for the „secondary decontamination" at the facilities of the medical evacuation chain (**PCHB-60-P** and **PCHP-60-P**). This system was based

C. Dishovsky et al. (eds.), Medical Treatment of Intoxications and Decontamination of Chemical Agents in the Area of Terrorist Attack, 153–163.

on subsequent use of two solutions, i.e. **Solution No 1:** 15 % (w/v) of sodium cresolate in 96 % ethyl alcohol, and **Solution No 2:** 20 % (w/v) of sodium benzene sulphochloroamide (chloramine B) in 82 % (v/v) ethyl alcohol in water with 11.6 % (w/v) of zinc(II)chloride.

After appearance of new agents with high percutaneous toxicity (V-agents) in the 1960s, it was necessary to test this two-solution system on the effectiveness in decontaminating V-agents. The chemical formula of VX was published in the early 1970s, but our suspicion on the possible structure of this agent led us to start R&D works on all problems of physical and medical protection (detection, decontamination, first aid, therapy, means of personal protection etc.) against V-agents using a model substance, i.e. the N,N-dimethyl analogue of VX, later depicted as medemo in the world literature. The author initiated this research in the Czechoslovak NBC-Defence R&D Establishment (now Military Technical Institute of Protection, Brno) with synthesis and determination of basic chemical, physical and toxic properties in the group of O-alkyl S-n-(N,N-dialkyl)aminoalkyl methylphosphonothiolates already in 1963 - 1964. To the first tasks belonged testing of impact of these agents on means of protection and laboratory decontamination available at that time, in order to assure workplace safety in R&D. The works on the problems of medical protection performed mainly at then Purkyně Medical Research Institute (now Faculty of Military Health Services, University of Defence, Brno) in Hradec Králové started with already standard compounds by toxicometrics in the late 1960s continuing with works on the first aid and experimental therapy later.

2. THE SEARCH FOR NEW MEANS OF SKIN DECONTAMINATION

The author started systematic research on the problems of personal decontamination in 1971. These works commenced with testing of efficiency of the available two-solution system as the matter of reference in any further developments. General toxicity testing was performed in mice, rats and guinea pigs, but mainly rats were used for testing of percutaneous toxicity and of decontamination efficiency. Some experiments were made also on guinea pigs and rabbits. For this reason, standard methods of percutaneous contamination and standard methods of decontamination for both liquid and solid decontaminants were developed in the first line. It was found that the best way to express the decontamination efficiency is the **Decontamination Index DI_{50}**. This expression corresponds to the ratio of LD_{50} p.c. (mg/kg) *with decontamination* (under standard conditions) to the LD_{50} p.c. (mg/kg) *without decontamination*. The higher is the value of DI_{50}, the higher is the efficiency. $DI_{50} = 1$ means no decontamination.

Testing the decontamination efficiency of the two-solution system with acceptable effectiveness in case of GB, GD and HD showed insufficient effectiveness in case of medemo (later confirmed also for actual VX). This was the main impetus to start extensive R&D of the quite new means for primary decontamination and of upgrading the means for secondary decontamination.

Our research followed three main directions:

- Decontamination solutions and emulsions
- Decontamination ointments, gels and pastes
- Decontamination adsorption and chemisorption powders.

3. MATERIALS AND METHODS

CHEMICAL AGENTS

Main types of supertoxic lethal chemical warfare agents, i.e. HD, GB, GD, medemo and VX, as described in [1] were used undiluted.

EXPERIMENTAL ANIMALS

Mice, albino rats, guinea pigs, and rabbits were used, as described in [1].

STANDARD METHOD OF PERCUTANEOUS CONTAMINATION

Undiluted agents were put on prior shaved skin on *dorsal thorax* in the form of small droplets using specially calibrated platinum wires or loops, as
in [1].

STANDARD METHOD OF DECONTAMINATION (DETOXIFICATION)

If it was not necessary to determine the influence of time between contamination and decontamination on the decontamination efficiency (significantly decreasing in time), the decontamination was provided uniformly 2 min after the animal had been intoxicated. The whole prior sheared area (i.e. 15 sq.cm) was treated:

- In case of solid, 0.3 g of powder was uniformly spread over the whole area. Then the powder was softly rubbed on the whole area for 30 seconds, using twisted sections of cotton wool 42 x 45 mm (together 0.50 g), held with tweezers, as described in [2].
- In case of liquid, standard wickled gauze tampons (about 0.330 g) held with tweezers were soaked with standard contents of liquid mean (1.6 ml) by dipping into solution and the whole area was wiped for 30 seconds. In case of two solutions, the second solution was applied 1 min after the first one.

EVALUATION OF EXPERIMENTS

Percutaneous toxicities (LD_{50}) both without and with decontamination were calculated statistically using the probit-numerical method by means of elaborated programme on the computer HEWLETT-PACKARD 9830-A. Individual doses were applied to the groups of 5 - 10 animals and at least 5 points with the lethality between 5 - 95 % were to be found on the lethality vs. dose curve.

4. RESULTS AND DISCUSSION

PERCUTANEOUS TOXICITIES OF TESTED CHEMICAL AGENTS

Percutaneous toxicities of supertoxic lethal chemical warfare agents in tested experimental animals are summarized in table 1.

Table 1. Percutaneous toxicity of tested agents LD_{50} p.c. mg/kg [2] ± standard deviation or (lower - upper confidence limit).

Agent	Albino rat	Guinea pig	Rabbit
HD	**15.8** ± 3.28	---	---
GB	**128** ± 84.9	**98.5** (63.5 - 153)	---
GD	**15.8** ± 4.23	**22.7** ± 4.98	**5.9** (3.70 - 9.30)
Medemo	**0.0630** ± 0.022	**0.81** (0.52 - 1.82)	**0.057** (0.041 - 0.80)
VX	**0.0132** ± 0.007	**0.25** (0.20 - 0.32)	**0.051** (0.021- 0.11)

DECONTAMINATION EFFICIENCY OF THE TWO-SOLUTION SYSTEM

The carefully determined decontamination efficiency of the above mentioned first post-WW-II generation of means based on the subsequent use of two solutions, i.e. IPB-60 (OZB and like) used as reference standard to compare all R&D results, achieved later, is shown in table 2.

Table 2. Decontamination efficiency of IPB (OZB, PCHB-60-P and PCHP-60-P) [2] (DI_{50}, albino rats).

Agent	HD	GB	GD	medemo	VX
DI_{50}	11.4	8.1	23.5	71.7	107

The decontamination efficiency of VX seems high on the first view. But it is quite opposite. Let us only compare the the toxicity data and compare them vs. contamination density on the skin surface in mg/sq.cm [2]. Moreover, the non-sufficient rate of the decontaminating reaction even diminishes the protective properties of clothing [1] in case of decontamination on man [3]. Also skin irritation (observed in over 20 % of tested persons) was among reasons to start new R&D.

UPGRADING THE TWO-SOLUTION SYSTEM

The two-solutions system where the subsequent use of both solutions was necessary for primary decontamination because of lacking knowledge which agent the person has been contaminated with, contains solutions, otherwise relatively effective when used alone in case of exact knowledge of the agent to be decontamined (Solution No 1 - GB, GD, Solution No 2 - HD, VX) which might occur exceptionally, rather at secondary decontamination. This was the reason of effort to upgrade this system by partial innovation of PCHB-60-P and PCHP-60-P via substituting Solution No 2 while retaining Solution No 1 unchanged due to its excellent efficiency against GB and GD and low irritancy and toxicity.

In these research works, the main stress was laid on such oxidating agents offering the possibility of being stable in solution, having a low toxicity and irritancy. Good results were obtained with hydroperoxy compounds of various structures, compounds with active iodine with combined micellar effects, such as polyvinylpyrrolidone and like and with other chlorinating agents. The most promising results (utilising *inter alia* just introduced industrial production of chlorinated derivatives of trichloroisocyanuric acid in Czechoslovakia), were obtained with upgraded Solution No 2, where 15 g of trichloroisocynuric acid and 10 g zinc(II)chloride was dissolved in 100 ml of acetone [7]. The decontamination efficiency is presented in table 3.

Table 3. Decontamination efficiency of the upgraded two-solution system [4] (No 1 as in IPB, No 2 trichloroisocyanuric acid in acetone).

Agent	HD	GB	GD	medemo	VX
DI_{50} (rats)	6.1	5.1	11.3	345	949

Even the toxicity profile is more advantageous, as seen in table 4.

Table 4. Toxicity of some decontamination solutions [4] (LD_{50} ml/kg in mice).

Solution	**i.m.**	**p.o.**
No 1 (IPB)	1.42 (0.94 - 2.10)	2.17 (1.68 - 2.65)
No 2 (IPB)	2.05 (1.60 - 2.57)	1.85 (1.36 - 2.85)
No 2 (TCICA)	6.73 (3.57 - 9.61)	2.03 (1.23 - 3.77)

Also the skin irritancy is significantly lower for the upgraded solution No 2 containing trichloroisocyanuric acid (TCICA). This decreases to only 40 per cent of cases, observed at the original Solution No 2, while only slight irritancy (light erythema) occurs and no cases of heavy skin corrosion can be observed.

DECONTAMINATION WITH THE NEW CHEMISORPTION MEAN

In the search for simple, cheap and universal means, we have tested up to a hundred various materials with sorption (chemisorption respectively) properties. Very promising results were obtained in a group of specially treated bentonites of domestic origin. The final solution was acidically treated montmorillonite (i.e. with active H-centre) enabling to manufacture a simple means for primary decontamination that we depicted according to the method of employment as „sorption-mechanical"[5], while it is actually based on chemisorption. For optimal results of decontamination, some rubbing of the fine powder on the skin surface is recommended. This material, known under acronym „DESPRACH" during development, was introduced into the new means for individual decontamination **IPB-80** [6] in the Czechoslovak Army and as the substantial part of the set in the new kit for the first aid **ZPJ-80** [6] in the Czechoslovak Civil Defence. For some details on this mean, see [2].

This means meets all principal requirements, i.e.:

- Effectiveness against all main types of CW agents, i.e. universality,
- Speed of decontamination effect,
- Skin non-irritancy,
- Non-aggressivity on clothing material,
- Simplicity of manipulation,
- Readiness to use,
- Repeated use,
- Low weight,
- Use within a wide temperature range,
- Mechanical resistance,
- Extreme stability on storage,
- Simplicity of manufacturing enabling mass production,
- Accessibility of raw materials,
- Extremely low costs.

This means is actually very simple for production and use [6]. The powder is placed in a hand-operated PE bottle (in the form of quadrangular prism) with small orifice for directed spreading of powder on the contaminated area. The bottleneck contains a screw cap. It is interesting that this R&D resulted in an atypically cheap innovation: The price of the new IPB-80 was less than one third as compared with the old IPB-60 introduced twenty years ago. The most important property - the decontamination efficiency, is shown in table 5.

Table 5. Decontamination efficiency of the new method for primary decontamination (DESPRACH) based on chemisorption [4] DI_{50} (decontamination 2 minutes after contamination).

Agent	HD	GB	GD	medemo	VX
DI_{50} (rats)	12.56	3.73	19.03	968	1200

RESEARCH ON UNIVERSAL DECONTAMINATION SOLUTIONS

Even if our means for primary decontamination possesses beside universality and high cost-effectiveness also other advantageous properties as required for the means of primary decontamination, it seems that the use of this principle might not be the best solution under *all* circumstances, mainly in the case of secondary decontamination at the facilities of the medical evacuation chain where the use of solutions (emulsions) for decontamination of wounded personnel is presumed to enhance productivity in preparing prior intoxicated patients (where beside skin, also decontamination of clothing, and of dressing materials is necessary) for further surgical interventions. But there are generally divergent views on this matter, which can be witnessed with the renown rescue systems that have introduced sorption means (based on bentonite) for the secondary decontamination recently, moreover with lower effectiveness as compared with the Czech IPB-80 (e.g. Swedish Rescue Services – *Räddnings verket*) [8].

We started the research on universal solutions already in the late 1970s. Such research implies necessary compromises among some contradictory requirements. During this research work, having learned lessons from previous works, we laid stress on crucial methodical points and requirements:

- *In vivo* experiments (standard experimental animals - albino rats, guinea pigs, rabbits),
- Standard method of contamination (preparation of skin, p.c. intoxication techniques),
- Standard method of decontamination (volume, materials, techniques),
- Standard testing compounds (HD, GB, GD, VX, medemo),
- Toxicity of decontaminants (mice, i.v., i.m., p.o.),
- Human skin irritancy (human volunteers),
- Impact on textile materials,
- Impact on other relevant materials (including environmental impact),
- Stability of solution (stability of components) on storage,
- Temperature range of use,
- Accessibility of components,
- Simple use,
- Repeated use,
- Technological problems
- Cost-effectiveness.

Because of the different chemical nature and thus also ability of principal groups of supertoxic CW agents to undergo basic types of reactions (electrophilic and nucleophilic), it is obvious, how difficult it is to find one simple universal reactant for agents with higher affinity to nucleophilic reactions (such as G-agents) and for agents possessing higher affinity to electrophilic reactions (such as HD and V-agents). When considering also requirements related to toxicity, irritancy and corrosion, not to speak of other desired military technical and economical properties, it is clear how difficult is to find a polyvalent solution based on one chemical principle, or how to find a compromise among all requirements. Nevertheless, our results have clearly shown that it is possible to find efficient chemical decontaminants among some nucleophilic agents (even with regard to toxicity and irritancy). Anyway, only principles offering high reaction rate (competing with the penetration rate) have chance to be utilised in practice.

Due to the limited extent of this paper, we mention only some positive results achieved in the orientation to two main directions i.e. to oxidative principles and to alcoholates in aprotic solvents. The research of **oxidative principles** (see e.g. [9]), involved the use of compounds with active chlorine, active iodine, peroxo-compounds, various mixtures of these compounds, as well as the use of oxidative agents with enhancing solubility and thus decontamination efficiency by adding detergents. It can be concluded that e.g. iodophores of various origin, although effective on VX, were generally unsufficiently effective on GD. Similar results were achieved in case of various compositions of chloro-derivatives of isocyanuric acid. On the other hand, some promising results were achieved combining sodium peroxide with a suitable mixture of commercially accessible detergents, as shown in one example in table 6 [4, 9].

Table 6. Decontamination efficiency (DI_{50}) of the sodium hydroperoxide emulsion with detergents.

Agent	HD	GB	GD	medemo	VX
DI_{50} (rats)	3.9	3.9	22.6	4115	2660

This emulsion contains 12.5 per cent (w/v) of sodium peroxide, 0.5 per cent (w/v) of Slovanion, 0.5 per cent (w/v) of of Slovafol and 0.5 per cent (w/v) of Alkon in water.

This idea has been utilized in elaboration of similar mixture of alkaline hydroperoxide with foam-generating detergents recently [10] but with a lower efficiency (tested only on GD - DI_{50} 12.9). It is obvious that such emulsion cannot be used for primary decontamination in the first line due to a very limited stability (in the order of hours only) and also due to unsufficient effectiveness against HD. Such emulsion has thus limited utility. It could be used as a means of choice in stationary facilities to utilise its excellent effectiveness against V agents.

Another, by no doubt very promising direction represent **alkaline alcoholate solutions** in a very cautiously selected aprotic solvent, even with regard to skin irritancy and corrosivity (See e.g. [11]). As the main constitutive components, we suggested alkaline ethoxyalcoholate, monoethanolamine and ethylalcohol. Among aprotic solvents, dimethylsulphoxide (DMSO), dimethylformamide (DMF), acetonitrile, and acetone were tested. It was shown, that sodium alcoholates posses generally higher decontamination efficiency than potassium alcoholates. Potassium alcoholates are also slightly more toxic. Sodium alcoholates prepared from metallic Na have similar toxicity and effectiveness like alcoholates prepared from sodium hydroxide. A final assessment was performed according to decontamination efficiency, skin irritancy, toxicity, safety, stability, technological parameters and cost-effectiveness. In the suggestion for innovation of means for secondary decontamination (1983), all these points were considered. As for the decontamination efficiency, see example in table 7. Toxicities of all tested solutions as well as their skin irritancy are very low (see table 8).

Table 7. Decontamination efficiency of alcoholates in aprotic solvents.

Agent	HD	GB	GD	medemo	VX
DI_{50} (rats)	17.7	> 6.8	57	1790	1280

These values correspond to the composition: 10 % (v/v) MEA, 43 % (v/v) ethanol, 45 % (v/v) Me_2CO + 2 % (w/v) Na.

Table 8. Toxicity of new decontamination solutions (LD_{50} ml/kg, mice).

Solution	i.m.	i.v.	p.o.
Na – H_2O_2	> 20.0	1.43	6.58
Na alcoholate			
- with DMSO	12.13	1.12	9.55
- with acetone	18.92	0.90	7.77

Note: skin irritancy of these solutions is generally very low in humans

5. CONCLUSIONS

As a result of our R&D, a new means for primary decontamination of human skin and adjacent parts of clothing in the variations for both armed forces (IPB-80) and civil protection (ZPJ-80) has been introduced recently and a cheap and efficient mood of partial upgrading of the means for secondary decontamination (PCHB-60-P and PCHP-60-P) has been suggested. Achieved results on universal solution (emulsion) for decontamination of human skin including adjacent parts of clothing (or dressing in case of wounded personnel) have shown the possible use of sodium peroxide emulsion of detergents as a mean of choice with limited stability and extremely high detoxification efficiency against VX but low efficiency against HD. The most promising results with generally extremely high decontamination efficiency against all tested agents and low toxicity and skin irritancy were achieved with sodium alcoholate solutions (made from sodium hydroxide) containing monoethanolamine, ethylalcohol and selected aprotic solvent (such as dimethylsulphoxide, dimethylformamide, acetonitrile, acetone and like).

REFERENCES

[1] Matoušek J.: Protective properties of standard combat clothing against skin penetration of supertoxic lethal chemical warfare agents. In: Sohns T., Voicu V. A. (Eds.): *NBC Risks: Current Capabilities and Future Perspectives for Protection.* Kluwe Academic Publishers, Dordrecht – Boston – London, 1999, pp 303-310.

[2] Matoušek J.: Sorption-mechanical principle in skin decontamination. In: Sohns T., Voicu V.A. (Eds.): *NBC Risks: Current Capabilities and Future Perspectives for Protection.* Kluwer Academic Publishers, Dordrecht – Boston - London 1999, pp 265-269.

[3] Matoušek J.: Problem of decontamination of clothing on man. *6th International Sympos. on Protection against CBW Agents.* Stockholm 1998. *Proceedings, vol. 1*, NDRC, Dept. of NBC Defence Umea 1998, pp. 255-256.

[4] Matoušek J.: Means for decontamination of supertoxic lethal chemicals on human skin. *Symposium on Nuclear, Biological and Chemical Threats in the 21st Century, NBC-2000,* Helsinki-Espoo 2000. *Symposium Proceedings, Research Report No 75,* University of Jyväskylä 2000, pp. 216-221.

[5] Matoušek J.: Universal mean for primary decontamination on the body surface (in Czech). CS Pat.1403/T (1974).

[6] Matoušek J., Hodbod˘ J., Šebestík M.: Sorption-mechanical mean for decontamination of CW agents for multiple use (in Czech). CS Pat. 1931/T (1981).

[7] Matoušek J., Elsnerová I.: Liquid mean for decontamination of CW agents on the body surface (in Czech). CS Pat. 1409/T (1974).

[8] SEDAB AB, Sweden : Prospect materials in NBC Protection (1998).

[9] Matoušek J.: Universal solution for decontamination of CW agents on body surface based on oxidation principle (in Czech). CS Pat. 1951/T (1981).

[10] Cabal J., Kassa J., Severa J.: A comparison of the decontamination efficacy of foam-making blends based on cationic and nonionic tensides against organophosphorus compounds determined *in vitro* and *in vivo. Human & Experimental Toxicology* **22,** 507-514 (2003).

[11] Matoušek, J.: Alcoholate solution for decontamination of CW on body surface (in Czech). CS Pat. 1907/T (1981).

Chapter 19

HARMFUL CHEMICAL-INDUSTRY INCIDENTS EFFECTS PROGNOSIS SYSTEM OF THE TERRITORIAL CENTER FOR EMERGENCY MEDICINE: SUPPLY OF INFORMATION AND ANALYSIS DATA

Alexander A. Kolyada
Department of medicine of catastrophes
MEDICAL ACADEMY OF POSTGRADUATE EDUCATION, Zaporizhia, Ukraine

Abstract: More than 3,930 thousands of tons of hazardous chemical substances are stored or used in production activities of the 78 industrial plants that operate in the Zaporozhia region. 1061, 28 thousands of people live in this area. These production sites can provoke the interest of terrorist groups due to the big number of people working there and become targets for terrorist acts. To improve the system for reaction to extraordinary situations related to chemically dangerous productions and increase the technological security of chemical production complexes, it is necessary to develop and introduce a complex of measures.

A method on the prognosis of consequences from hazardous chemical substances to be used during failures in industrial plants and during transportation in order to predict eventual consequences was developed.

Keywords: automated control system; contamination; hazardous chemical substances; medicine of catastrophes; sanitary-epidemiology state

The creation in Ukraine of a state service for medicine of catastrophes stipulated the necessity for this service to be in a state of everyday readiness to implement the functions laid on it, above all things the operative management of extraordinary situations.

The nowadays system for supply and receipt of information does not always answer the current needs. In such conditions, it is impossible to consider the information on the concerned subject of management as being objective, since it is coming in too late. Unreliable untimely information sometimes results in errors during the decision-making administrative process and in extraordinary situations can cost the life or be the instrument for worsening the condition of victims. As a result, the necessity becomes clear to develop and

C. Dishovsky et al. (eds.), Medical Treatment of Intoxications and Decontamination of Chemical Agents in the Area of Terrorist Attack, 165–170.

introduce in the state service for medicine of catastrophes an automated control system. The system should be able to provide the operative management by liquidating the medical consequences in emergency situations regardless of the place and time of their happening.

It is well known that for the creation of informative products which is the first step in the process of management, it is necessary to have resources [2].

In the frames of the basic goal to protect the population from the harmful effects of failures, there are a number of measures to be implemented for the protection of and for the diminishing the consequences of failure on personnel and local inhabitants in areas with dangerous chemical production industry. More than 3,930 thousands of tons of hazardous chemical substances are stored or used in production activities of the 78 industrial plants that operate in the Zaporozhia region. These substances include over 0,9 thousands of tons of chlorine, over 1,831 thousands of tons of ammonia and about 1,101 thousands of tons of other hazardous chemicals.

Depending on the diffusion of chemical danger the production sites have been graded as follows:

- First degree are 3 sites (more than 3 thousand peoples live in the area of the possible chemical damage);
- Second degree are 2 objects (0,3 to 3 thousand people live in the areas of the possible chemical damage);
- Third degree are 53 sites (0,1 to 0,3 thousand people live in the areas of possible chemical damage);
- Fourth degree are 20 sites (in the area of the possible chemical damage live less than 0,1 thousand people).

These production sites can provoke the interest of terrorist groups due to the large number of people working there and become targets for terrorist acts.

The safety during operation of chemically dangerous production sites depends on many factors: physical and chemical properties of the raw material, character of technological process, construction and reliability of equipment, terms of storage and transporting of chemical compounds, state of control and measuring devices and facilities for automation, efficiency of facilities for emergency protection. In addition, safety of production, use, storage and transportation of extremely poisonous compounds depend in a considerable degree on the level of organization of prophylactic work, timeliness and internals of preventive-maintenance, repair works, preparedness and practical skills of personnel, system for supervision, status of emergency protection hardware.

For the improvement of the system for counteracting emergency situations due to chemically dangerous productions, and the increase of the technological security of chemical production complexes, it is necessary to provide the development and the introduction of a set of measures:

- identify production sites that represent danger of chemical origin, declare the status of their safety, insurance and registration according to the requirements of the Law of Ukraine "About potentially dangerous productions";
- ensure that managers and businessmen take responsibility for the emergency situations arisen from dangerous productions and are punished according to the existing legislation;
- situate the production sites rationally and keep them distant from populated areas, taking into account the potential danger in emergency situations;
- use of construction material with protective properties, establish systems with automated sensors of control, introduce systems for automatic disconnecting on the incident site of the electric power, fuel, gas, and systems for fire-alarm and the like;
- in relation to the detailed study of the sanitary-epidemiologic status of all regions should be taken into account the development of highly efficient immune preparations and facilities for immediate prophylaxis, facilities to carry out mass sanitation of the population, disinfection and fumigation of different objects and localities;

Warning for failures in plants, including such resulting from terrorist acts, requires development and introduction of new technique and technologies' standards, which should respond to the world up-to-date safety requirements. Diminishing the risk of failures demands the development of new production methods, the introduction of the newest technologies, technological processes which substantially diminish the risk of failures.

The complex of measures for prevention and minimization of the effects of technological emergency situations in chemically dangerous production sites should include:

- Application of state-of-the-art chemical technologies in the aim of preventing the occurrence of industrial failures and protecting the people and environment;
- Creation in the plants, which operate with highly poisonous matters, of a local systems for warning upon exposure or damage of the environment and notification of the staff and the population, that lives in the area of the possible chemical infection;
- Preparation for early prognostication of areas of chemical contamination of environment objects;
- Accumulate the necessary amount of facilities for individual and collective protection of the workers and the people which live in areas of possible chemical damage.

A method for prognosis of consequences after spread of harmful chemical substances due to failures in industrial plants or during transportation was developed. The method can be used in industrial plants and in different kinds of transportation facilities: pipeline, motor, river, railway and overseas.

It can also be used for long-term emergency prognosis at failures in chemically dangerous productions and transportation facilities, as well as to determine the degree of chemical danger for populated areas and for the administrative-territorial units [1].

However, for the receipt and treatment of information about the status of hazardous productions in general and especially in cases of failure due to terrorist acts, it is necessary to establish the necessary informative ducting in advance. Under informative channel we mean the way information transfer is done, regardless of the facilities, which includes the system of information transfer, system of transport of information and eventual system or systems for reception of information. Directions of information flow can change depending on active jobs.

For an effective and timely reaction it is necessary to send informative streams in certain directions: the emergency object is the center for urgent medical care and rescues services. In addition, for a reliable prognosis of failure consequences, it is necessary to have informative streams in the direction from the meteorological service to the center of urgent medical care. The establishment of active sensors of surveillance covering large parts of the chemically dangerous production territory and including this data in the network of the informative and analytic system will be of great assistance in extraordinary situations. That kind of information received from the sensors that monitor potentially dangerous productions should be accessible by the Government service of medicine of catastrophes for subsequent treatment and prognostication of situations, both operative and long-term. For such prognostication modern information technologies should be available.

The system for warning in extraordinary situations should be part of a multifunction multilevel informative system, that is envisaged to include all potentially dangerous production sites of the country. The lower level of this system consists of separate elements located on the potentially dangerous sites or of observation posts near potentially dangerous natural objects. From all the elements of the lower information level system data will flow into the regional level system, which is the concern of the region. Information from the regional system will be sent to the situational center. Such is generally speaking the chart of functioning of the system for warning in extraordinary situations.

We will discuss the main features of the offered system. At the lower level the state of all potentially dangerous sites - both natural and technogenic will be controlled. The periodicity and the facilities for carrying out the control over these sites is the concern of special expert subsystems, established on the basis of technical documents (for technical sites) or of data from the Ministry for ecological protection (for natural objects). For example, the sites of the first category should be equipped with special checking for the reliability of the security systems located directly on the site and continuous supervision of the status.

The system can function in one of three modes: foreground mode, prognosis mode and warning mode. In the foreground mode information is given out on the sites' and region status for the present day. In the prognosis mode - the system shows how the establishment of a new enterprise or the closing of an existing one will influence the status of safety of the region. In the warning mode - the system registers any weakening of the level of security in the region. By design, the system should be able to foresee the possible consequences of subsequent development of the situation, to offer a bank of recommendations and give out a list of possible measures for prevention in extraordinary situations or for diminishing the harmful effects, if such situations are inevitable.

This system can be realized on the basis of computing engineering supported by the proper software. Personal computers at the working places should be interconnected locally and regionally, covering the whole territory of the region. It follows that it is wise to create the expertly-deliberative system for prognosis of possible consequences of extraordinary situations at territorial level. Modern information technologies using the properties of computers allow to create a unique up-diffused database, with possibilities for transparent (within the limits of the system) passage of information both forward and in reverse direction with confirmation of receipt by the addressee; easy and frequent addition of operative information about the extraordinary situation that acts to the knot; easy processing of information in any indexes and receipt of reports on such indexes for subsequent treatment during planning of measures for urgent medical care during emergency situations; timely automatic creation of reports: unique as well as standard and their presentation to higher levels; watching of the medical activities for liquidation of the effects of extraordinary situations and timely receipt of additional or final information from the units of lower level; decrease of the expenses for long-distance telephone service from the budget of Central District Hospital due to optimization of informative streams within the limits of the system; electronic signature of documents which are passed in the system [5].

REFERENCES

1. Order of MINISTRY emergency of measures of Ukraine, Ministry of agricultural policy, Ministry of economy and Ministry of ecology and natural resources of Ukraine «About claim of Method of prognostication of consequences of outpouring (extrass) of hazardous chemical substances at failures on industrial objects and transport» from March, 27, 2001 № 73/82/64/122
2. Voloshyn V. A., Terentyeva A. V., Terentyev A. V. Modern directions of development of the automated control systems by government service of medicine of catastrophes // Announcer of social hygiene and organization of health protection of Ukraine. - 2000. - №3. - P. 37-40

3. Voloshyn V. A., Terentyeva A. V., Rogach Sh. M., Zagoruyko N. L. Information providing of Government service of medicine of catastrophes of // is the Ukrainian magazine of extreme medicine of the name of G. O. Mozhayev. - 2001. - №2. - P. 16-18
4. Kartysh A. P., Roschin G. G., Voloshyn V. A., Guryev S. A. et al. Organization of work of the information's analytic system of Ministry of health protection of Ukraine on the questions of extraordinary situations // The Practical manual. - Ministry of health protection of Ukraine, USPC of DM. Kiev. - 2002. - 102 p.
5. Kolyada A. A., Kochin I. V., Sydorenko P. I. Problems of prognostication of possible sanitary losses from extraordinary situations with the use of the expertly-deliberative computer system of // The Organizational, medical-pharmaceutical and methodical aspects of medicine of catastrophes. Materials of allukrainian scientifically-practical conference. Ternopil. - 2005. - P. 30-31
6. Olijnyk P. V., Yevstratyev Ye. Ye. Mathematical design of organization of providing of population by medications in the conditions of extraordinary situations // The Pharmaceutical magazine. - 2003. - №1. - P. 23-28

Chapter 20

PURIFICATION OF DRINKING WATER FROM $^{134,137}Cs$, $^{89,90}Sr$, ^{60}Co AND ^{129}I

Rashid A. Khaydarov, Renat R. Khaydarov
Institute of Nuclear Physics, 702132, Ulugbek, Tashkent, Uzbekistan

Abstract: The most probable and dangerous radionuclides which can appear in waste water of atomic power stations after accidents or terrorist acts are $^{134,137}Cs$, ^{90}Sr, ^{129}I. Moreover, the radionuclides ^{137}Cs, 90Sr and ^{60}Co are also most attractive for terrorists to use in dirty bombs. As a result of the disasters the radionuclides can contaminate drinking water. Generally, , tap water is decontaminated at water supply stations. Nevertheless it is very important to have sorbents and household drinking water filters in stocks to prevent panic of populace and consequences of emergency cases at the water supply stations.

As the concentration of Na^+, K^+, Ca^{2+}, Mg^{2+}, Cl^- ions in potable water can have the value of 100 - 500 mg/l and the concentration of radionuclides is very low, the sorbents must have high selectivity and distribution coefficient of Cs^+, Sr^{2+}, Co^{2+}, I^- ions. Unfortunately most of known resins cannot be used for these purposes.

The created technology of producing cationic and anionic exchange fibroid sorbents on the base of polyacrylonitrilic fibers is described in the paper. Chemical characteristics of the sorbent and filters were investigated. The static exchange capacity is 3-4 meq/g for cationic sorbents and 1-2 meq/g for anionic sorbents. The removal coefficient of listed radionuclides from drinking water by the filter is 10^2 - 10^3.

After the Chernobyl accident the Academy of Sciences and KGB of Ukraine tested the household filters with fibroid sorbents and ten thousands filters were produced in the Institute of Nuclear Physics of Uzbekistan and given to Ukraine through "Isotope" Corp. (USSR). The experience of using the filters for purification of drinking water from radionuclides in Chernobyl region is described in the paper.

High technical and chemical characteristics of the sorbents have been confirmed by tests in certified labs of Germany, USA, Korea, India and Russia.

Keywords: anion-exchange sorbents; cation-exchange sorbents; Chernobyl; drinking water; exchange capacity; Radionuclide; terrorist act

C. Dishovsky et al. (eds.), Medical Treatment of Intoxications and Decontamination of Chemical Agents in the Area of Terrorist Attack, 171–181.

1. INTRODUCTION

The most probable and dangerous radionuclides which can appear in waste water of atomic power stations after accidents or terrorist acts are $^{134,137}Cs$, ^{90}Sr, ^{129}I. Moreover, the radionuclides ^{137}Cs, ^{90}Sr and ^{60}Co are also most attractive for terrorists to use in dirty bombs. As a result of the disasters the radionuclides can contaminate drinking water. Generally tap water is decontaminated at water supply stations. Nevertheless it is very important to have sorbents and household drinking water filters in stocks to prevent panic of populace and consequences of emergency cases at the water supply stations.

As the concentration of Na^+, K^+, Ca^{2+}, Mg^{2+}, Cl^- ions in potable water usually have the value of 100 - 500 mg/l and the concentration of radionuclides is very low, the sorbents must have high selectivity and distribution coefficient for Cs^+, Sr^{2+}, Co^{2+}, I^- ions. Moreover these sorbents must be very cheap because the filters cannot operate for a long time due to accumulation of radioactivity. Unfortunately most of known resins cannot be used in the filters for these purposes.

The object of this work was creating the fibroid sorbents and filters which meet requirements given above, demonstration of their advantages and disadvantages on the base of the experience of use in the Chernobyl region.

An advantage of the fibroid ion-exchange sorbents over resin is in high rate of a sorbtion process, effective regeneration and small value of pressure drop of the sorbent layer for purified water (Kragten 1978, Zverev 1981, Khaydarov 2003). The specific surface of the fibroid sorbents is $(2 - 3)\cdot 10^4$ $m^2/$ kg, i.e. about 10^2 times greater than that of the resin (10^2 $m^2/$ kg). Owing to that fact the rate of the sorbtion process on the fibroid sorbents is much greater than that on the resin.

The main drawback of the fibroid sorbents is their very low specific weight value of 50 - 200 kg/m^3.

2. MATERIALS AND METHODS

Polyacrylonitrile (PAN) cloth with surface density of 1.0 kg/m^2 and thickness of 10 mm was utilized as the raw material for making ion–exchange sorbents.

1-10% solutions of NaOH, 5 - 40% solution of hydrazine hydrate $NH_2NH_2\cdot H_2O$ and 0.5-5% solutions of polyethylenimine $(-NHCH_2CH_2-)_x[-(CH_2CH_2NH_2)CH_2CH_2-]_y$ were used for treatment of PAN clothes to make ion-exchange sorbents.

The 0.001M $CuCl_2$ solution labelled by ^{64}Cu and $K_2Cr_2O_7$ solution (pH 2) labelled by ^{51}Cr were used to find out the best technology of making cation- and anion- exchange sorbents, respectively. Radionuclides ^{64}Cu and ^{51}Cr were made by irradiating $CuCl_2$ and $K_2Cr_2O_7$ at the nuclear reactor of the Institute of Nuclear Physics (Tashkent, Uzbekistan). Ge(Li) detector with a resolution of about 1.9 keV at 1.33 MeV and the 4096-channel multichannel analyzer were used to detect γ-quantum from radionuclides. Areas under γ-peaks of radionuclides ^{64}Cu (half-life $T_{1/2}$ is equal to 12.8 h, energy of the γ-peak E_γ is equal to 0.511 MeV) and ^{51}Cr (half-life $T_{1/2}$ is equal to 27.72 d, energy of the γ-peak E_γ is equal to 0.320 MeV) were measured to calculate the amount of Cu and Cr respectively.

Other radionuclides used in the investigations of sorbents characteristics are given in Table 1.

Table 1. Radionuclides used as labels.

Elements	Radionuclides	$T_{1/2}$	E_γ , MeV
M-^{32}P	^{32}P	14.3 d	E_β=1.7
Cr(VI)	^{51}Cr	27.73d	0.320
Co(II)	^{60}Co	5.27 y	1.17, 1.33
Ni(II)	^{65}Ni	2.5 h	1.480
Cu(II)	^{64}Cu	12.7 h	0.511
Zn(II)	^{65}Zn	244.1 d	1.115
Br(I)	^{82}Br	35.3 h	0.776
Sr(II)	^{89}Sr	50.5d	0.909
M-^{99}Mo+^{99m}Tc	^{99}Mo+^{99m}Tc	66 h (6.0h)	0.140
Cd(II)	^{115}Cd	53.5 h	0.336
Sb(II)	^{124}Sb	60.2 d	1.691
M-^{131}I	^{131}I	8.04 d	0.364
Cs(I)	^{134}Cs	2.07y	0.605

The exchange capacity Q, meq/g, was calculated by Equation (1):

$$Q = (A_0 - A_e)/(A_0 - A_B) \cdot B/W, \qquad (1)$$

where B is amount of carrier, meq; W is weight of exchanger, g; A_0 is a count rate of the original solution, A_e is a count rate of the solution at equilibrium, A_B is a background count.

The distribution coefficient K_d and the percent adsorption P were calculated by Equations 2,3:

$$K_d = ((A_0 - A_B)/(A_e - A_B) - 1)\cdot V/W, \quad (2)$$

$$P = 100\cdot(1 - (A_e - A_B)/(A_0 - A_B)), \quad (3)$$

where V is total volume of the solution, ml.

The sorption processes of ions from water in dynamic conditions were studied by using columns with diameter of 12 mm; the weight of the sorbents was 1 g.

3. RESULTS AND DISCUSSION

The chemical process of making cation-exchange sorbents consists of treatment PAN cloth by solutions of $NH_2NH_2\cdot H_2O$ and NaOH. Following chemical transformations take place after treatment by hydrazine hydrate:
- formation of linear frames, and then formation of intermolecular chemical bonds.

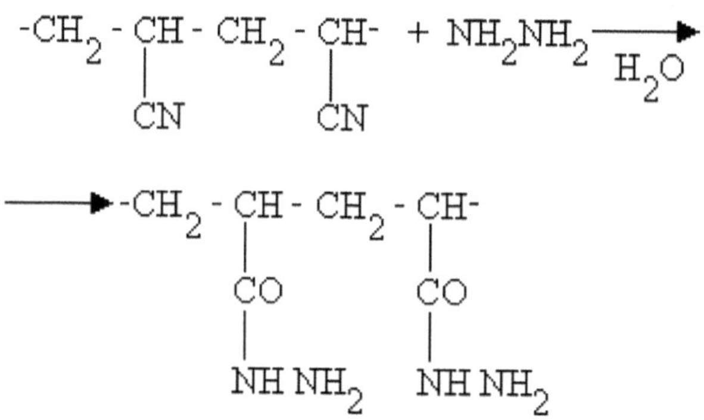

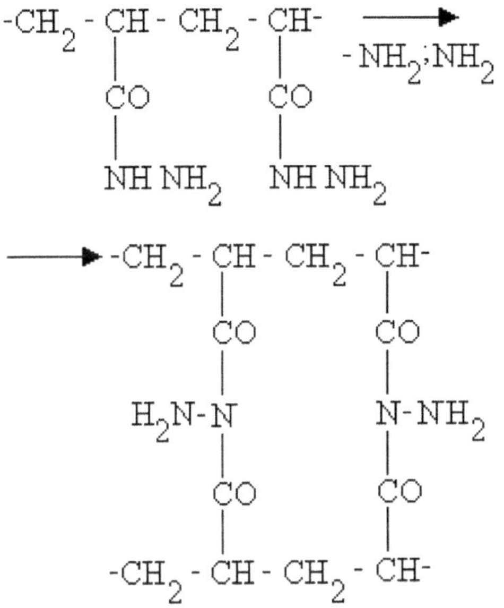

After saponification of the nitric groups by alkali carboxyl groups are formed:

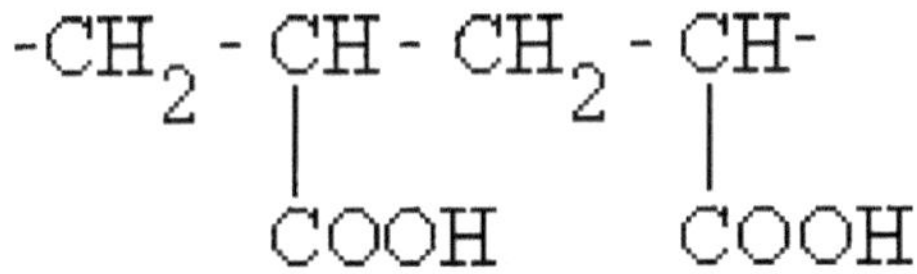

Kinetics of saponification of the PAN fibers treated by 5 - 40% solution of $NH_2NH_2{\cdot}H_2O$ at 40-95 °C were studied in the range of NaOH solution concentration from 1 to 10% at 25 - 70 °C. For example, the kinetics of saponification by 5% NaOH solution of samples treated by 20% $NH_2NH_2{\cdot}H_2O$ solution at 70 °C are given in Fig.1.

Increasing the temperature of treating solution and duration of treatment cause filter capacity increase, but strength of the fibers degrades. Experiments have shown that optimum condition is treatment by 20% $NH_2NH_2{\cdot}H_2O$ solution at 70 °C during 30 minutes and by 5% solution of NaOH at 25-30°C during 1hour. Exchange capacity (Cu^{2+}) of the sorbents is 3.5 - 4.0 meq/g.

Anion-exchange sorbents are made by treatment of cation - exchange filters in H-form by water solution of polyethylenimine. Amine groups attach to carboxy groups by electrostatic forces. Kinetics of anion-exchange groups' formation at concentrations of polyethylenimine from 0.5 to 5% and the temperature range from 20° to 70°C were studied. Fig. 2 demonstrates kinetics curve at 25°C and concentration of polyethylenimine 1% and Fig. 3 shows dependence of exchange capacities against the concentration of polyethylenimine at 25°C and treatment time of 1 hours.

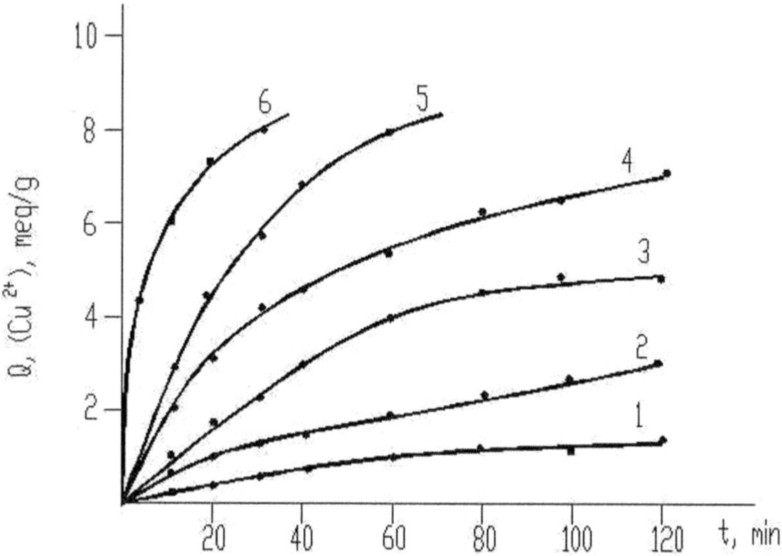

Figure 1. Kinetics of saponification of the fibers in 5% NaOH solution at 25°C (1), 30°C (2), 40°C (3), 50°C (4), 70°C (5) and 90°C (6).

The treatment of the cation-exchange sorbents by 1% solution of polyethylenimine at 25°C during 1 hour was selected as the optimal condition for the anion-exchange sorbents production. Capacity (Cr^{6+}) of the sorbents is 2.0 meq/g.

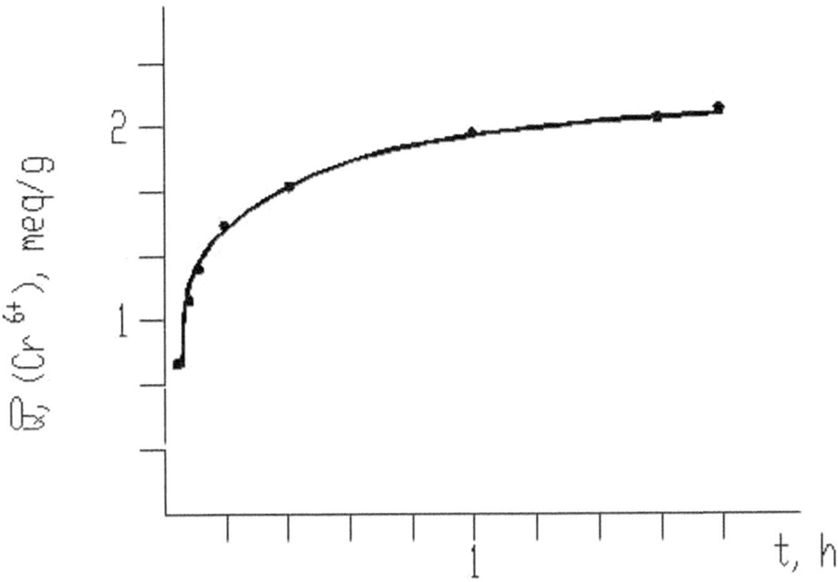

Figure 2. Kinetics of anion-exchange groups formation at 25°C in 1% polyethylenimine solution.

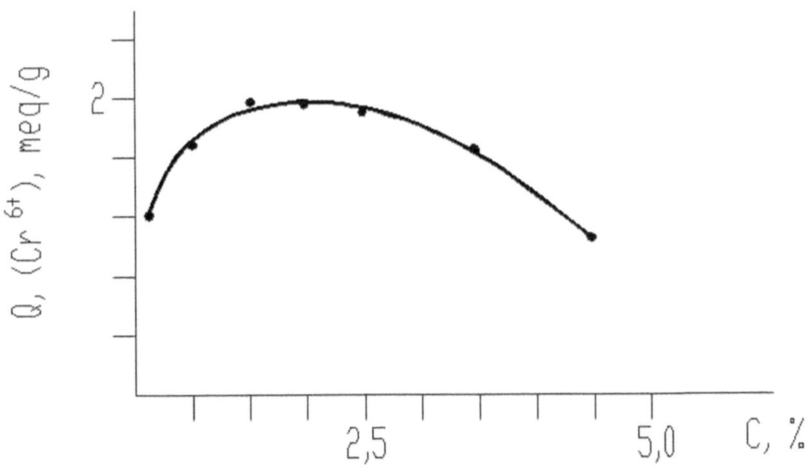

Figure 3. Dependence of exchange capacities against the concentration of polyethylenimine at 25°C.

Removing radionuclides (^{51}Cr, ^{60}Co, ^{65Ni}, ^{64}Cu, ^{65Zn}, ^{89}Sr, ^{115}Cd, ^{124}Sb, ^{131}I, ^{134}Cs, etc.) from water was studied. Dependence of the distribution coefficient K_d for various ions against pH of the solutions is presented in Table 2.

Table 2. Distribution coefficient K_d (mL/g) for various ions and organic substances (C_0 = 10 mg/L, V=50mL, W=0.5 g).

Elements	Exchanger	pH of solutions									
		1	2	3	4	5	6	7	8	9	10
Co(II)	Cationic	3000	2600	2300	2000	1700	1000	126	138	150	160
Ni(II)		125	600	870	920	990	750	430	510	780	1000
Cu(II)		140	400	600	480	400	560	650	560	460	340
Zn(II)		230	2000	4000	5000	4000	1900	1700	1400	900	800
Sr(II)		11	25	45	100	300	1000	1900	8000	6000	900
Cd(II)		980	830	680	520	380	240	97	75	46	17
Sb(II)		260	190	150	130	120	120	115	90	70	35
Cs(I)			100	200	900	1900	3200	4000	4000	1500	11
Cr(VI)		200	150	100							
I(I)	Anionic		3100	2800	2600	2300	2100	1900	500	150	

The specific behavior of K_d of Co(II), Ni(II) and Cu(II) is explained by dependence of the relation between the M^{n+} form and the hydrolyzed forms in the solution with pH (J. Kragten 1978).

The influence of additional foreign cations Na^+ , K^+, Ca^{2+}, Mg^{2+} on the adsorption of different ions at pH = 7 is presented in Table 3.

Table 3. Influence of additional foreign cations Na^+ and K^+ on the distribution coefficient K_d (mL/g) of different ions at pH=7.

Radionuclides	Na^+		K^+		$Na^+ + K^+$
	10 mg/L	100 mg/L	10 mg/L	100 mg/L	100+ 100 mg/L
^{60}Co(II)	130	140	130	140	140
^{90}Sr(II)	510	300	520	420	500
^{134}Cs(I)	4000	4000	4000	4000	4000
Radionuclides	Ca^{2+}		Mg^{2+}		$Ca^{2+} + Mg^{2+}$
	10 mg/L	100 mg/L	10 mg/L	100 mg/L	100+ 100 mg/L
^{60}Co(II)	130	130	130	130	130
^{90}Sr(II)	510	500	520	500	500
^{134}Cs(I)	4000	3700	4000	4000	4000

The produced cationic and anionic exchange fibroid sorbents on the base of polyacrylonitrilic fiber were used to produce household water filters. The cloths of cationic and anionic exchange sorbents with mass of 50 g each one were reeled on a perforated cylinder with diameter of 10 mm and length of 15 mm. This cartridge could be easily replaced. Initial tap water passed through the sorbents from the perforated cylinder. Water purified from radionuclides overflowed through the nozzle of the filter. Dependence of the removal coefficient of ^{134}Cs, ^{90}Sr, ^{129}I, ^{60}Co for the filter against pH at the concentration of Cs^{+}, Sr^{2+}, Co^{2+} and I^{-} of 0.1 mg/L are given in Fig. 4-5. These experiments have shown that the removal coefficients of listed radionuclides from water by the filter is 10^2 - 10^3 at a pH range of 6-8.

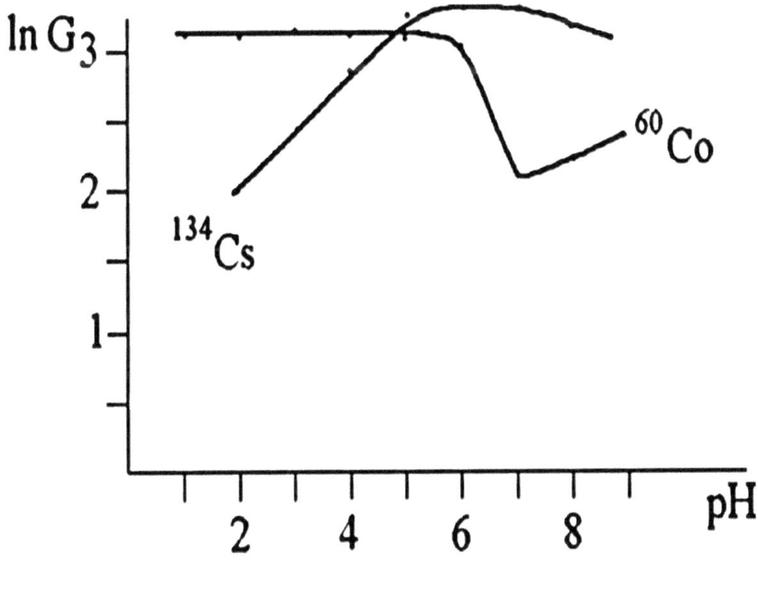

Figure 4. Dependence of the removal coefficient G of ^{134}Cs and ^{60}Co for the filter against pH at the concentration of Cs^{+} and Co^{2+} ions of 0.1 mg/L.

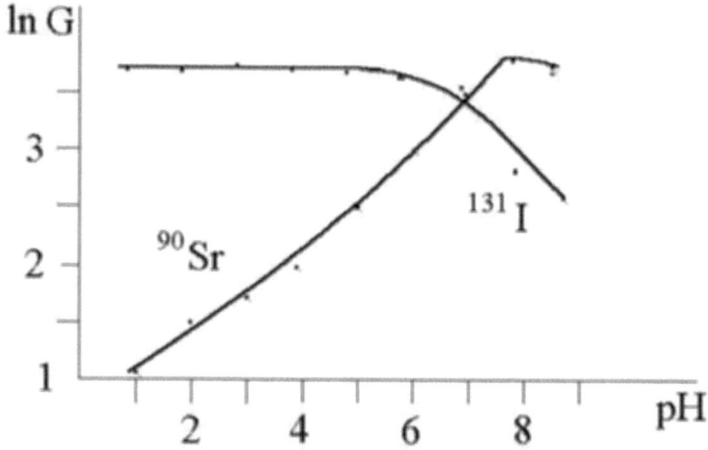

Figure 5. Dependence of the removal coefficient of ^{90}Sr and ^{129}I for the filter against pH at the concentration of Sr^{2+} and I^- ions of 0.1 mg/L.

The filters after Chernobyl accident were tested by Academy of Sciences and KGB of Ukraine on real water contaminated by radionucledes. Tables 4 and 5 illustrate typical results of water purification process and process of radioactivity accumulation by the filters, respectively.

Table 4. Concentration of radionuclides (nCi/L) in Chernobyl's water after household filter.

Radionuclides	Initial concentration, nCi/L	Volume of water passed through filter, m^3			
		0.5	1.0	5.0	10.0
^{137}Cs	1.5	0.12	0.15	0.14	0.16
^{90}Sr	0.4	0.06	0.07	0.07	0.07

Table 5. Accumulation of radioactivity (nCi) of the household filters after purification of Chernobyl's water.

Radionuclides	Initial concentration, nCi/L	Volume of water passed through filter, m^3			
		0.1	1.0	5.0	10.0
$^{137}Cs + ^{90}Sr$	1.5 + 0.4	$2 \cdot 10^2$	$2 \cdot 10^3$	$1 \cdot 10^4$	$2 \cdot 10^4$

These experiments have confirmed that the filters have removal coefficients of radionuclides for water from Chernobyl of 10^2 - 10^3 · Thus, the water meets the requirements for drinking water after passing the filter. However, the high efficiency of the filter causes problems due to accumulation of radioactivity by filters, increasing the difficulty of controlling radioactivity at homes and the necessity of removal of radioactive wastes. In accordance with "The Main Sanitary Specifications" OSP-72/87 of Russia the solid wastes are radioactive if their specific activity is greater than $2 \cdot 10^{-6}$ Ci/kg for sources of β-particles and $1 \cdot 10^{-7}$ gram-equivalent of Ra/kg for sources of γ-quantum. Thus the solution was accepted that the exploitation resource of the household filters in the Chernobyl region mayt not exceed 50 L and after termination of life the cartridges of the filters could be considered as a "cold" waste and may be replaced.

Due to the very low price of the cartridges (less than $1) it was possible to prevent panic of populace and solve the problem. Ten thousands filters were produced in the Institute of Nuclear Physics of Uzbekistan and sent to Ukraine through "Isotope" Corp. (USSR).

4. CONCLUSIONS

The described results of the investigation shows that created chemically modified PAN fibroid filters have satisfactory adsorption characteristics. The capacity of the cation-exchange sorbents is 3-4 meq/g (Cu^{2+}) and that of the anion - exchange is 1-2 meq/g (Cr^{6+}). The cation- and anion-exchange filters are selective for removing radionuclides $^{134,137}Cs$, ^{90}Sr, ^{60}Co and ^{129}I in presence of Na^+, K^+, Ca^{2+}, Mg^{2+}, Cl^- ions in water at concentrations up to 500 mg/L. The removal coefficient of the listed radionuclides from drinking water by the filters is 10^2 - 10^3.

Experience of using these filters with the removable cartridges in Chernobyl regions has shown that they have been able to solve problems conducted with radioactive contamination of potable water: preventing panic of populace, preventing the water contamination in the distance between the water supply station and consumers, preventing consequences of emergency cases at the water supply stations, supply with decontaminated water at remote settlements, etc.

REFERENCES

1. J. Kragten, *Atlas of Metal-Ligand Equilibria in Aqueous Solution*, Ellis Horwood Ltd., Chichester, 1978.
2. M.P. Zverev, *Fibroid chemosorbents*, Chemistry, Moscow, 1981.
3. R.A. Khaydarov, O.U. Gapurova, R.R. Khaydarov, S.Y. Cho, Fibroid sorbents for purification of air, waste water and drinking water, Proceedings of the First International Conference on Environmental Researche and Assessments, Bucharest, Romania, March 23-27, 2003, p.153-163.

Chapter 21

RESEARCH OF DECONTAMINATION EFFICIENCY OF CONTAMINATED CAMOUFLAGE CLOTHING WITH APPLIED METHODS AND MEANS IN REPUBLIC OF BULGARIA

Tsvetan Popov[1], George Popov[2]
[1]State Agency for Civil Protection, Bulgaria
[2]Kingston Environmental Laboratory, USA

Abstract: Decontamination ability of the new developed and applied camouflage clothing is researched via evaporation and treatment with air and mixture of evaporated water and ammonia. Kinetics of the processes depending on the temperature and time is determined. Optimal conditions for efficient decontamination and apply in technical means for decontamination of the supplied materials in Republic of Bulgaria are specified.

Keywords: decontamination efficiency, mustard agent, desorption, camouflage clothing

Obligations of the countries ensuing from the convention for the prohibition of chemical weapons do not reduce the actuality of the protection. Terrorist acts during last decades with use of chemical and biological weapons enforce not only elaboration of new more efficient protection means, but permanent development and improvement of methods and means for consequence management.

In spite of numerous researches devoted to improve methods and means for special treatment efficiency of decontamination in a number of instances does not correspond to the new requirements. An explanation for this from one side is the higher level of toxicity of the new chemical warfare agents small quantities of which could be very dangerous to humans and, from the other side, the process of decontamination is dependent both on methods and decontamination ability and specific properties of the contaminated object.

Decontamination ability of the new developed and applied camouflage clothing was not fully researched. Contamination with chemical

C. Dishovsky et al. (eds.), Medical Treatment of Intoxications and Decontamination of Chemical Agents in the Area of Terrorist Attack, 183–189.

warfare agents and decontamination with chemically active substances could have a significant effect on basic characteristics of camouflage garment.

The aim of this research was to determine decontamination ability of the new camouflage garment of the Bulgarian Army with applied methods and means and to access the influence of the decontamination on physicomechanical characteristics of the camouflage garment.

As objects of research samples of camouflage textile material with dimensions 20x50 mm were used. For contamination of the samples we used mustard agent (β,β'-dichloroethyl sulphide) with a purity of 95,6 %. Contamination was accomplished with a special dropper (figure 1) resulting in a weight of the droplets of $0,5.10^{-6}$ kg and a density on the test material of 5.10^{-3} kg/m^2.

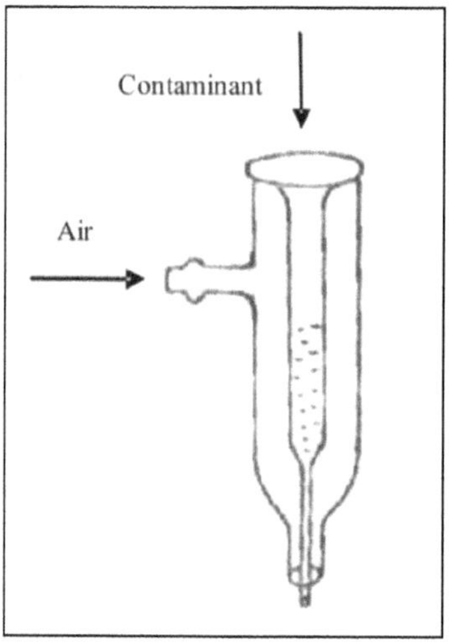

Figure 1. Dropper for contamination.

Desorption with hot air and decontamination with air and mixture of evaporated ammonia and water was realized in special glass desorber (figure 2).

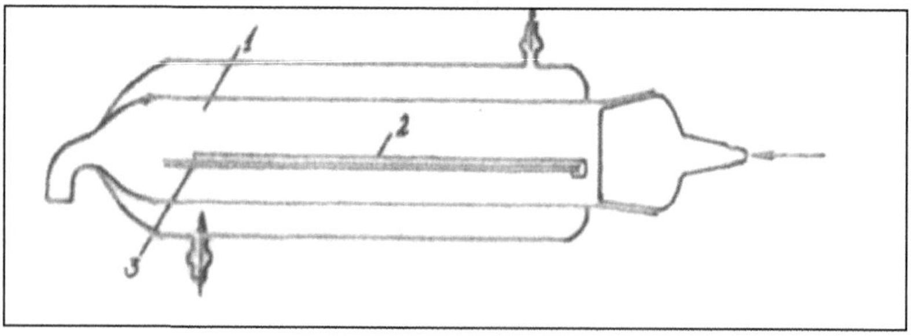

Figure 2. Desorber 1. Corpus 2. Sample 3. Glass Mount.

Extraction of mustard agent from the samples was realized with ethyl alcohol. The remains of the toxic substance in samples were determined with the spectrophotometer "SPECOL" according to the thymolphtalein reaction [1, 2] with preliminarily made calibration curve.

Decontamination of the samples was provided via:

- Evaporation with hot air;
- Treatment with air and mixture of evaporated ammonia and water.

Evaporation of the mustard agent was performed in desorbers (fig. 2) mounted in thermal chamber (fig. 3).

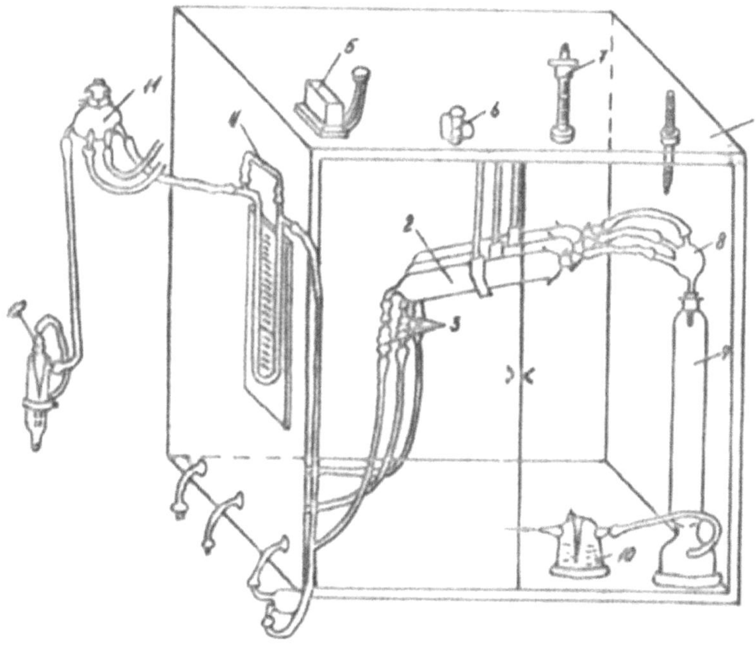

Figure 3. Thermal chamber. 1. Thermal chamber 2. Desorber 3. Silica gel cartridge 4. Flow-meter 5. Thermo relay 6. Fan 7. Contact thermometer 8. Air distributor 9. Drying column 10. Air distributor 11. Water pump.

The aim of the research via evaporation was:

- Quantitative determination of evaporated and residual quantities mustard agent depending on temperature and time;
- Determination of the processes and specifying opportunities for natural decontamination and decontamination with hot air produced in decontamination machines AGV-3U and station for consequence management "Cristanini" – Italy.

Kinetics of evaporation of contaminated samples according to temperature is shown on fig. 4.

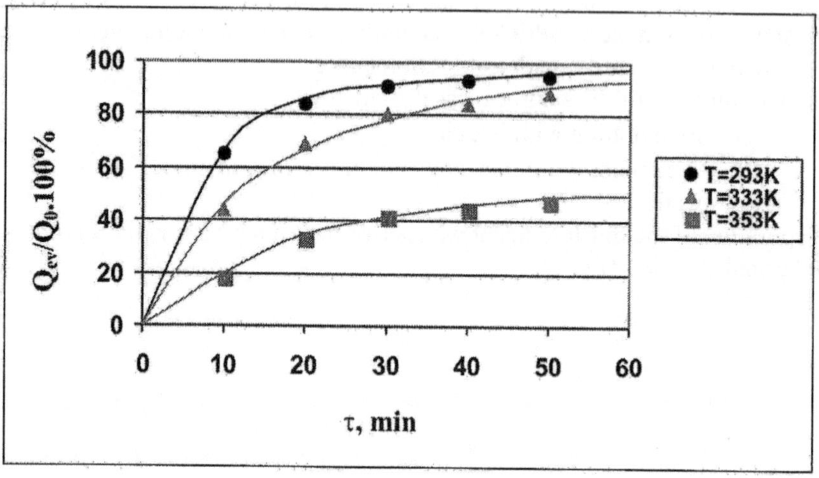

Figure 4. Kinetics of evaporation of mustard agent from camouflage garment according to the time and temperature. Exposition of contamination of the samples 3600 s. Velocity of airflow 5 l/min.

Analysis of the results shows that as a difference from evaporation of mustard agent from lacquer covers where kinetics of the process passes through three stages evaporation of mustard agent from textile material passes in two stages. An explanation for this is that in the moment of the experiment the whole quantity of mustard agent is penetrated into the textile material while on the lacquer covers not the whole amount of mustard agent is penetrated resulting in a greater velocity of evaporation. Evaporation in the first stage goes with higher velocity and after 20 minutes at high temperatures most of the mustard agent is vaporized (about 75-85%). In this stage the velocity of evaporation is very high because of the excess of the contamining substance and "bonds" of mustard agent with textile material which are mainly of a physicochemical nature. During the second stage evaporation of remained quantities of mustard agent bonded chemically to textile is started. For a long time insignificant quantities of mustard agent are evaporated. By increasing the temperature the velocity of evaporation regularly increases too and time needed for decontamination is shortened. At 20° C the velocity of evaporation is too slow and an insignificant amount of mustard agent is evaporated (about 40% for 50 min) within the period of investigation. Thus it is not reasonable to rely on natural decontamination of mustard agent via evaporation. At 60°C the velocity of evaporation increases considerably but decontamination is not complete within the time scaleof investigation, i.e., after 50 minutes the amount of evaporated mustard agent is ca. 90%. Stages of evaporation are well expressed in the curve at 80°C, where 10 minutes after starting the experiment more than 80% of the initial amount toxic substance is evaporated and around the 30th minute the evaporated mustard agent is about 94%. Around the 40th minute the amount of evaporated substance is 95% and

50 minutes after the start evaporation is about 96%. It can be seen that not the whole amount of mustard agent is evaporated and about 4-5% still remain in the samples "bonded" to textile material chemically. These chemical bonds are strong enough and could not be removed practically. It is known that mustard agent has a tendency to create hydrogen bonds [3] with substances containing double bonds, hydroxyl groups, etc.

Cotton textile material and used dyes for painting contain a lot of such bonds and groups and probably create hydrogen and other chemical bonds with mustard agent. According to reference [3] remained toxic substances "bonded" chemically samples do not exert influence on the physiological condition of the organism.

Decontamination of the treated samples with mixture of air and vaporized ammonia was conducted in a thermal chamber (fig. 3) with flows of purified dry air and evaporated ammonia through desorbers. Ammonia was produced form crystalline ammonium hydrogen carbonate. An evaporator filled with a 10% solution of hydrogen carbonate is heated to 80°C and subsequently assembled to the system for purified dry air.

The aim of the research of decontamination with air and mixture of evaporated ammonia and water was:

- Determination of the process-specifying kinetics of decontamination with mixture of air and evaported ammonia;
- Determination of the efficiency and conditions of decontamination with a mixture of air and evaporated ammonia of treated samples;

The kinetics of decontamination of the treated samples is shown on fig. 5.

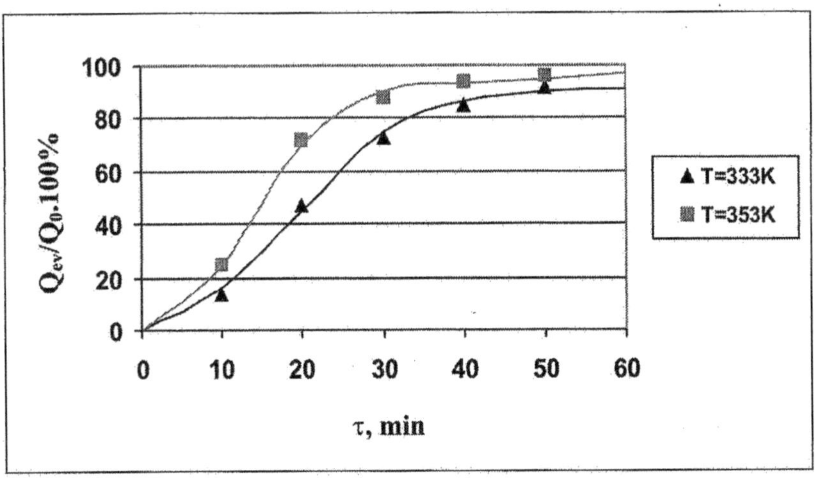

Figure 5. Kinetics of decontamination on camouflage coatings with air and mixture of evaporated ammonia and water depending on temperature. Exposition of contamination 3600 s..Velocity of the mixture of air and vaporized ammonia 5 l/min. Concentration of ammonia in flow – 3 mg/l.

Analysis of the results shows that as a difference from evaporation, decontamination with a mixture of air and evaporated ammonia has three stages. This could be explained with different components in the passing flow. While during evaporation only heated air passes through and the velocity of evaporation increases, in decontamination the flow is saturated with evaporated ammonia. In the first stage (up to 10 minutes after starting the experiment) the mixture moistens the samples and the velocity of decontamination is low. In the second stage, after moistening the samples and reaching the required temperature the velocity of the process increases high (for about 10 minutes) and 70-80% of the initial amount mustard agent is decontaminated. After the second stage (about 20th minute) starts the third stage. This stage is the most prolonged, but during this time insignificant amount of mustard agent is decontaminated (4-5%). This could be explained with the remained quantities of mustard agent, "bonded" to coating and paintings. In contrast with evaporation, decontamination is accomplished with chemical reactions according to the scheme:

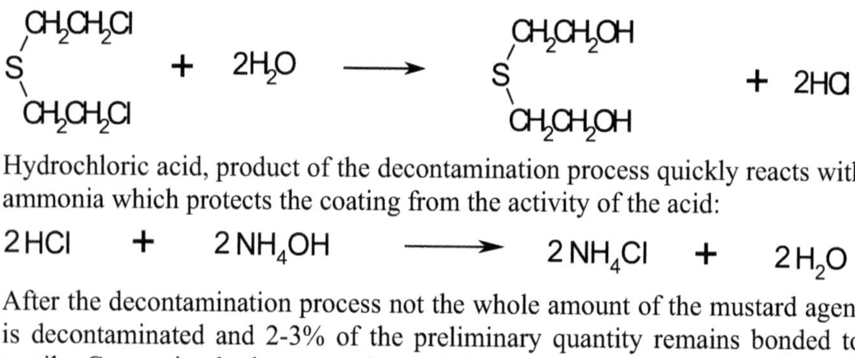

Hydrochloric acid, product of the decontamination process quickly reacts with ammonia which protects the coating from the activity of the acid:

$$2HCl + 2NH_4OH \longrightarrow 2NH_4Cl + 2H_2O$$

After the decontamination process not the whole amount of the mustard agent is decontaminated and 2-3% of the preliminary quantity remains bonded to textile. Comparing both evaporation and decontamination the temperature has a significant influence on the kinetics of the process. At of 60°C mustard agent is not fully decontaminated in the researched period of time.

Physicomechanical properties of the samples using both methods (strength of stretching and dividing) change about 3-4 % which is within the limits of permissible effect.

Testing separately samples using both methods at 100°C learns that that the velocity of the processes increases considerably and the time for treatment is shortened but the values of physicomechanical properties are reduced with 8-9%. That is a reason to consider that optimal conditions for treatment using both methods are 80°C during 50 minutes. These conditions could be created and maintained in means for decontamination and disinfection of devices and armament.

CONCLUSIONS

1. The kinetics of decontamination of the new camouflage coatings is determined via evaporation and treatment with air and a mixture of evaporated ammonia and water depending on temperature and time.
2. Both methods of treatment reached the required efficiency.
3. Optimal conditions for treatment of the coatings are specified – temperature 80°C, time for treatment – 50 min. Physicomechanical properties of garment are reduced after having been treating three times with 3-4%, which is within the limits of permissible influence.in the Republic of Bulgaria in accepted means for decontamination and disinfection of devices and armament.

REFERENCES

1. Franke Z., P. Frantz, "Chemistry of Poisonous Substances" Part I and II, Moscow 1973.
2. Adamovich I.S., N.I. Ivanov, "Basis of Deactivation, Decontamination and Disinfection", Moscow 1968.
3. Knunyantz I.L., "Chemical Warfare Agents", Moscow 1967.
4. Brasser, P., "Modeling the Chemical Protective Performance of NBC Clothing Material", Journal of Occupational and Environmental Hygiene, 2004.
5. Smith, J.M., and E. Stammers: Physical Transport Phenomena I [translation]. Delft: DUM, 1986.
6. NATO Army Armaments Group (NAAG), Land Group 7 (Joint NBC Defence): Operational Requirements,Technical Specification and Evaluation Criteria for NBC Protective Clothing (AEP-38). Brussels: NATO, 1998.

Chapter 22

ALLEVIATION OF TOXIC IMPACT OF CHEMICAL AGENTS ON HUMAN ORGANISM

Tetyana Vasylyeva,[1,2], Yekaterina Duka [1], Mykola Kharytonov [3]

1) Department of Pediatrics, Dnepropetrovsk State Medical Academy, 9a Dzerzhinskogo , Dnepropetrovsk, 49027, Ukraine,
2) Pediatrics Department Texas Tech University Health Science Center in Amarillo, 1400 Coulter str., Amarillo, TX 79106, USA
3) Ecotoxicology laboratory, Dnepropetrovsk State Agrarian University, Voroshilov st.25, Dnepropetrovsk, 49600, Ukraine

Abstract: Industrial toxicants as risk factors for development of dysadaptation syndrome among children population dwelling in unfavorable environmental conditions were under investigation. The response to environmental toxicants in children population was polymodal. The higher concentration of chemicals in environment, the more children with anomalous physiological reaction were found. Children living in polluted areas more often had hypertrophy of tonsils, increased lymphatic nodes and liver size and dismorphic features. Negative influence of toxicants took the form of secondary immune-deficiency state, which was expressed by repeated respiratory infections, etc. Children dwelling in environmentally unfavorable areas showed significant decrease in immunity. Due to the accumulation of heavy metals in humans, effective rehabilitation treatment, which includes pectin-vitamin tablets, natural adaptogenes and antioxidants, have been tested and proposed.

Keywords: air pollution, heavy metals, antidotes, and human health.

1. INTRODUCTION

The health of the population is closely related to the environmental situation in industrial areas. Heavy metals were the priority contaminators in large industrial centers of Ukraine for a long period of time [1]. They penetrated into the environment inside industrial wastes, and then circulate in trophic chains and reside in tissues of plants, animals, and humans. Our previous analyses showed that soil Zn deficiencies

C. Dishovsky et al. (eds.), Medical Treatment of Intoxications and Decontamination of Chemical Agents in the Area of Terrorist Attack, 191–201.

reduce reproduction of the population and increase the total level of children's mortality and morbidity. There were positive correlations between concentration of Co in soils and malignant neoplasm, diabetes and infant mortality, and between the concentration of Mn in soils and infant mortality [2]. Changes in contaminated areas occur faster than the adaptation process of human organisms. Contaminators penetrate into organisms and change physiological processes, decrease the resistance toward infective agents and mutagens. Due to that background various diseases can occur with higher frequency. Some diseases could be prevented by the early recognition of the ecological desadaptation syndrome and by taking measures to raise the resistance of organism. Previous data showed that the reaction of children population is polymodal - there were three groups with strong, moderate and weak reactions toward environmental pollutant. That individual reaction mostly depended on genetic stability of the organism. However, the higher the concentrations of toxicants were in the environment the more children with anomalous of reaction were found. Children dwelling in polluted areas had hypertrophy of tonsils, increased lymphatic nodes, and liver size. This finding related to an increase of the strain in functioning of protection barriers of organism. It's known that the state of growth development of children could independently characterize the effects of environmental contaminate on health. Children dwelling in environmentally unfavorable conditions had increased heights with worsening of the state of health and decreased functional reserves of organism. In polluted areas signs of dimorphism were registered more often. Toxic effect of xenobiotics disturbed metabolism in the examined group of children. For example, the amino acids spectrum of the urine was changed dramatically compared with healthy children from control area (districts with limited levels of environmental pollutants). Direct connection between some growth indices, functional state of cardiovascular and respiratory systems and the level of excretion of amino acids in the urine was established. The negative influence of xenobiotics on the state of the morphofunctional, biochemical and genetic substratum of organism led to strain in functioning of protection barriers. It took the form of a secondary immune-deficit state, which was expressed by repeated respiratory infection, etc. Children living in unfavorable areas had decreased cellular and humoral immunities [3]. Therefore, we proposed efficient rehabilitation measures of decontamination treatment, which include humus preparation and pectin-vitamins tablet, etc [4,5]. Humus substances increase non-specific resistance of the human organism. They have adaptogenous, antimutafacient, antitoxic, immune-corrective and other effects that may be employed to correct pathologies caused by the impact of unfavorable environmental factors. As a result of the treatment it was observed that normalization of immune parameters correlated with a decrease of cytogenetic pathologies and clinicical improvement [4,5]. Our previouse study showed the ability of sodium huminate to increase resistance of organisms toward the influence of harmful environmental factors. The effect was related to the influence on enzymatic reactions (for example, suppression of activity of 5 - oxy - tryptophane decarboxylase, taking part in the synthesis of serotonine) [4]. Treatment by pectin tablets resulted in normalization of immune parameters that correlated with the decrease in cytogenic pathologies and total clinical improvement. Children in condition of subcompensation (?) also could consider taking medicine which improve their immune response (penthoxyly, sodium nukleinatis and eleutherococcus). In the condition of decompensation of the immune status of a child it is necessary to include extract of thymus-like substances (timogeni, timalini, T-activini). Timely and differential prescription of immuno-rehabilitation complexes allows to stop the process of desadaptation of immune, endocrine, respiratory and cardiovascular systems due to diminish influence of the environmental pollution toxic effect.

2. MATERIALS AND METHODS

We investigated pollution of some soils and snow samples in the Dnieperside Region, and established the concentration of heavy metals in humans (in urine, saliva, hair, nails) by atomic absorption spectroscopy. Studies of the state of health of children dwelling in areas with different pollution levels included questionnaires for parents, a special chart used in clinical history and physical examination, assessment of functional tests of respiratory (peak flow meter) and cardiovascular systems (measurement of pulse, blood pressure, ECG), immune response related tests (levels of subclasses of immunoglobulins, T-and B-cells quantitative analysis), evaluation of some anthropometrics parameters (height and weight) and their comparison with clinical and functional characteristics. All measurement has been made with the use of standard techniques. Examination of 164 children dwelling in areas with different pollutant loads was performed (60 in district A, 60 in district B and 44 in district C). Statistics included correlation analysis and F and t – tests.

3. RESULTS AND DISCUSSION

3.1. Toxic industrial compounds in Dnepropetrovsk City

Dnepropetrovsk Region has numerous toxic industrial industries. The main pollutant sources are facilities of metallurgy, power generation, mining, as well as chemical and petrochemical industries. These managed emission sources are controlled better than the non-managed sources including quarries, waste depositories, and ash heaps. Up to 80 % of the total industrial emissions in the southeast part of Ukraine are related to enterprises of mining-metallurgical complex. For example, in the Dniepropetrovsk region the concentration of mining and metallurgical production is 7–10 times higher than the national average. Production and processing of manganese ore is 100% of the national total, iron ore is 86%, cast iron production is 50%, and the production of steel is 47% of the national total. There are 57 metallurgical companies in the Prydneprovsky Region. The metallurgical sector accounts for 64% of total regional emissions (about 530.000 tons per year) and is especially a source of the following substances (in thousand tons per year): dust – 90, SO_2 – 50, CO -- 370, NO_2 – 20, VOC (volatile organic compounds) – 2. Eleven thermal power stations cause tremendous air pollution (total amount was about 250 thousand tons per year, including dust – 80, SO_2 –120, CO – 15, NO2 -- 50). The wastes of all kinds from chemical companies are very different and are toxic for the environment, but the total amount of emissions into the atmosphere is about 2000 tons per year. The greatest emissions took place in Dnepropetrovsk because of the high concentration of environmentally dangerous industries. At the same time the tendency of some industrial gases to emission decrease was observed for the last years in Dnepropetrovsk [6, 7]. We investigated pollution of some soils in the districts of the Dnepropetrovsk with different technogenic load of heavy metals (A, B, C). (Fig. 1.1).

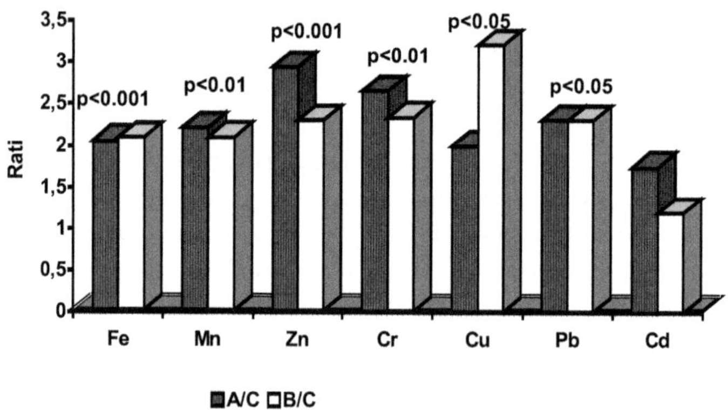

Figure 1. Soil Pollution Comparative Evaluation.

In particular, the concentration of Mn and Zn was double the amount in the soils of the pollution regions than in the control zone. The concentration of Cd was 1.8 times higher and the Pb concentration was 2.3 times higher times in the contaminated area than in clean territory. Snow samples were taken to reflect air pollution (Table1.1).

Table 1. Snow pollution with trace elements in different districts, mkmol/l.

Element	Districts		
	A	B	C
Cu	9,4±3,56	6,68±3,42	5,11±1,6
Mn	152,39±7,85	55,5±3,58	30,23±0,18
Pb	11,26±3,69	3,74±0,61	3,38±0,48

Thus it was shown that snow water samples in the Districts A&B are polluted too.

3.2. Comparative evaluations of some clinical and laboratory data in children dwelling in different environmental conditions

The analyses of questionnaires shows that some parents (49%) had a low level of ecological knowledge and they did not use any prevention of the environmental dysadaptation syndrome. Meanwhile, it is very important to diagnose this syndrome and to prevent that various diseases occur on caused by this background. It is also very important to increase the educational effort toward familiarizing the population with the consequences of the influence of environmental pollutants on human organism.

The higher concentration of a chemical in the environment, the more children with anomalous reactions could be found. Children living in polluted areas had hypertrophy of tonsils, enlarged lymphatic nodes, and enlarged livers (Tabl. 2).

Table 2. Comparative evaluation of some clinical data in children dwelling in different environmental conditions (%).

Clinical data	Pollution areas (A& B)	Control (C)
P		
Hypertrophy of tonsils	52	33.3
<0.05		
Enlarged lymphatic nodes	70	46.6
<0.05		
"Absorb" cardiac tone	20	6
<0.05		
Enlarged liver	18	3.3
<0.05		
Carious	22	23.3
>0.05		

Anthropometrical parameters and their comparison with clinical and functional characteristics show that children living in environmentally unfavorable conditions are characterized by worsening states of health and a decrease of functional activity. Thus, the average pulse rate was 93.9 ± 1.4 per minute in contamination zone and 84.3 ± 1.61 per minute in the controls. The functional ability of the respiratory system also decreases in a statistically significant way.

Heavy metals possess mutagenic effect. The assessment of mutagenic effect was assessed through evaluation of genes recessive mutation frequency in the micronucleus in oral mucous cells of children living at the tested areas (Tabl. 3).

Table 3. Micronuclei in oral mucous cells of children living in polluted and control areas (%).

Data	Polluted areas	Control
Number of children	41	24
Micronucleus in oral mucous	0.078±0.031	0.042±0.001

Some toxic effects of heavy metals pertain to a change of amino acid contents in urine. It is established that accumulation of metals in hair and nails and secretions through urine and saliva are definitely correlated. It was obviously due to enzyme features as well as the level of environmental pollution.

Heavy metals also have a profound impact upon the immune system of human organism. The level of T-lymphocytes in the blood of children dwelling in the pollution areas was 1,7+0,1 (10^{12}/L) against 2,93+0,09 (10^{12} /L) in control zone, the level of pre-B-lymphocytes was 0,29+0,05 (10^{12} /L) against 0,65+0,063 (10^{12} /L) (Fig. 2).

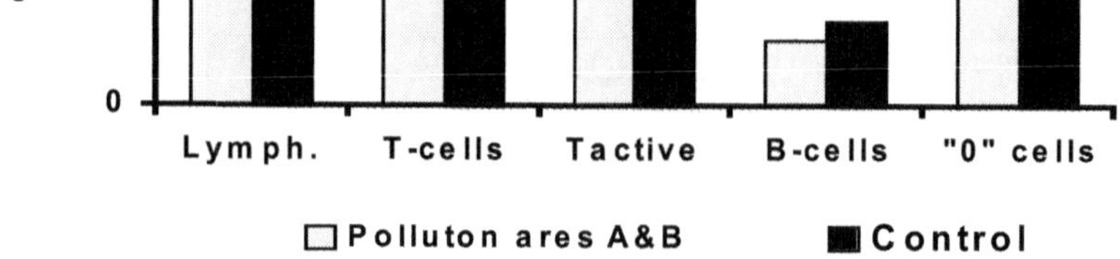

Figure 2. Cell immunity in different industrial regions.

There were also different levels of immunoglobulins (Ig) in saliva of children from contaminated and control zones (Tabl. 4).

Table 4. The level of Ig in saliva of children from different regions.

Data	SIgA (mkg/ml)	IgA (mkg/ml)	IgG (mkg/ml)
Pollution areas A&B(?)	574±52	66±6	65± 6
Control	406±33	64±8	156±23
P	<0.01	>0.05	<0.01

3.3. Rehabilitation Measures

The following preparations were used to increase children's' organism resistance in unfavorable environmental conditions: carotic oil - one teaspoon each day during one month (as antioxidant), enterosorbent - one teaspoon 3 times a day one hour before meal during 7 days (as antidote) and sanation of nasopharynx by eubiotics (A-bakterin) and sodium huminate. Compared with the control group the respiratory morbidity of the children who received the rehabilitation measures was 1.5 times less. The indices of systolic volume, minute volume, and coefficient of effectiveness of blood circulation reliably improved. 100 % children had peak speed of expiration increased, on average by 31±12%.

Children who had recurrent bronchitis and received rehabilitation measurement significantly increased the levels of immunoglobulins (Ig) (Fig. 3).

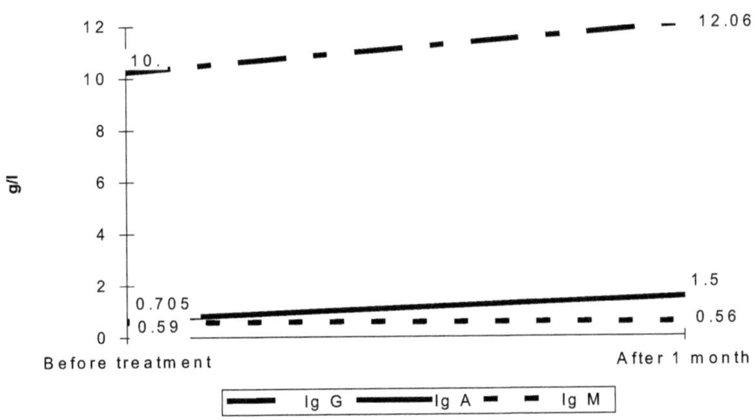

Figure 3. Changing of Ig levels in the blood serum under influence of rehabilitation treatment.

In case of immune depression the rehabilitation stimulated the main indices. Children who had an increased the number of T-lymphocytes at the beginning of study, mainly by increasing number of receptor subpopulation, reliably increased the number of active cells and teophyllin-resistant lymphocytes. The activity of phagocytes and the level of IgA increased. The patients with low content of pre-B-lymphocytes (on average 5.2±0.7%) also had a significant increase of this cells pool (on average up to 13±3%, $p<0.05$). Besides, under the influence of rehabilitation treatment the quantity of micronuclei in cells of the mouth cavity mucous decreased from 0.064±0.006 % to 0.047± 0.006 %. This fact shows the minimizing of genotoxic activity. Girls have had a decline from 0.06±0.05% to 0.05±0.06%, boys - from 0.066±0.006% to 0.047±0.006% (table 5).

Table 5. Changes in the quantity of the cells with micronuclei in the mouth cavity mucous of the children with recurrent bronchitis under influence of humus substances. (M±m)/

Investigated	Before treatment, %	After treatment, %	P
Girls	0.06±0.05	0.05±0.06	>0.05
Boys	0.066±0.006	0.047±0.006	<0.05
Total	0.064±0.006	0.047±0.006	<0.05

Taking into consideration the accumulation of heavy metals in children's organisms an effort has been taken to supplement rehabilitation treatment with detoxicant remedies, in particular, enterosorbentics. A good protective effect was observed with pectin-vitamin tablets. Pectin, the main component of the remedy, binds heavy metal ions, which had accumulated in the organism due to the environmental pollution (from polluted air, water and soil) and promotes their removal from the organism. Pectin has colloidal, bioactive, haemostatic, detoxicant, antacid and antiseptic properties. Pectin substances are high-molecular derivative of carbohydrates and a component of cell walls and medium plates of plant tissues. Pectin, which we used, was extracted from sugar beet. The effect of pectin on the organism is connected with its colloidal properties. For example, pectin adsorbs acetate of lead better than activated coal does. It has been argued that the products of pectin decay in combination with other compounds have therapeutic effects, rather than pectin itself. Galacturonic acid as a product of decay, which decreases influence of toxic substances, is considered to be of great importance. The binding effect is explained by its ability to complex formation. After getting into alimentary tract, pectin forms gels. The swollen mass of pectin dehydrates the alimentary tract and takes up the toxins while moving through the intestines. The remedy was taken at a dose of 1 gram per day (4 tablets) in the course of four weeks. After taking the remedy the content of lead in saliva was reduced to trace amounts, and the number of children with lead registered in their hair had diminished from 53.8 % to 33.3 %. Accumulation of copper positively reduced, whereas the concentration of zinc and manganese were halved. The amount of cadmium also somewhat decreased.

Thus, by fixation of heavy metals in the alimentary tract, pectin preparations decrease the quantity of xenobiotics, which pass through the organism and are fixed in tissues. (Fig. 4)

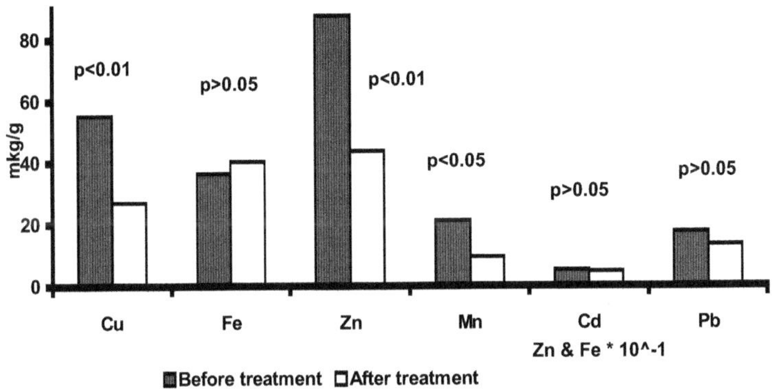

Figure 4. Concentration of heavy metals in the hair of children with recurrent bronchitis before and after usage of the pectin-vitamin tablets in therapeutics.

Under the influence of pectin-vitamin pills the quantity of micronucleus cells of mouth cavity mucous decreased from 0.060+0.007% to 0.041+0.007%. This fact shows minimizing of genotoxic activity. Girls have had a decline from 0.05% to 0.04%, and boys - from 0.064 to 0.05%. Similar changes, and even more pronounced, have occurred in leukocytes of the peripheral blood. (Table 6)

Table 6. Changes in the quantity of the cells with micronuclei in the peripheral blood of children under influence of pectin-vitamin tablets. (M±m).

Investigated	Before treatment, %	After treatment, %	P
Girls	0.034±0.005	0.005±0.0006	<0.001
Boys	0.037±0.006	0.01±0.0008	<0.001
Total	0.036±0.0032	0.009±0.0008	<0.001

The children dwelling in industrial areas could benefit from spending some vocational time in the recreational zones well far from large industrial areas and supplement their rehabilitation course with natural adaptogenes that increase the resistance of the organism (extractum eleutherococci, aralia), antioxidants, vitamins, antibiotics. Thus, remediation measures including pectin substances are very important in ecopathological conditions.

4. CONCLUDING REMARKS

A child's organism is very sensitive to environmental pollutants. The reaction could involve changes in various protective barriers including the immune system. Antioxidants, immune corrective therapy, eubiotics and antidotes play a substantial role in health protection. Pectin-vitamin tablets alone also have a significant effect. After taking the remedy supplements, the contents of heavy metal ions in the organism decreased. The contents of lead in saliva reduced to trace amounts, and the number of children with lead registered in hair diminished. Accumulation of copper essentially diminishes, the concentration of zinc and magnesium could be cut in half, and the amount of cadmium also somewhat decreased. Thus, fixing heavy metals on the level of gastrointestinal tract, pectin preparations decrease the quantity of xenobiotics which pass through the organism and are fixed in tissues. Under the influence of pectin-vitamin tablets the quantity of micronucleus in cells of mouth cavity mucus decreases. Similar changes, and even more clearcut, were ascertained in leukocytes of the peripheral blood. This fact shows a dcrease in genotoxics activity. Improvement of the environmental situation as well as protection from toxicants of sensitive groups of the population is an important step in order to enhance the health of future generations.

Acknowledgments

Credit is due to people who were involved in this work including, A.I.Gorovaya, T.Jaroshevskaya, M. V. Shyrkina , and M. I. Lukashenko.

REFERENCES

[1] Trakhenberg I.M., Kolesnikov V.S., Lookovenko V.P., 1994. Heavy metals in the environment. Minsk. 283 p.

[2] Vasylyeva T.L., Duka Y.D., Ogorodny V.V., Kharitonov N., 1997. Relation between Trace Elements Concentration in Soils and the Health of Humans in Ukraine. Fourth International Conference on the Biogeochemistry of Trace Elements, pp.51-52.

[3] Gorovaja A., Vassilieva T.. 1995, Cytogenetic aspects of mutagenic impact on biota and human beings from heavy metals of industrial regions' environment and rehabilitation effects of natural adaptogens. Third International Conference on the Biogeochemistry of Trace Elements, c.2. Paris, France

[4] Vasylyeva T., Duka E., Kharitonov N., 1995, Relation between polluted soils and health of population and rehabilitation measures, International Workshop on Technologies for Decontamination and Remediation of Soils Polluted by Chemical products, Paris -France

[5] Vasylyeva T., Duka E, Kharitonov N., 1996. Rehabilitation effect of humus substances in ecopathological conditions 10th International Peat Congress, Bremen-Germany

[6] Kharytonov M., Gritsan , N., and Anisimova,L. Environmental problems connected with air pollution in the industrial regions of Ukraine. In: Global atmospheric change and its impact on regional air quality. Barnes, I. (Ed). NATO Science Series, IV: Earth and environmental sciences. Kluwer Academic Publishers, 2002b; 16: 215-222.

[7] Babiy, A.,P., Kharytonov, M., M., Gritsan, N., P., Connection between emissions and concentrations of atmospheric pollutants. Melas D. and Syrakov D. (eds.), Air Pollution Processes in Regional Scale, *NATO Science Series, IV: Earth and environmental sciences.* 2003, Kluwer Academic Publishers. Printed in the Netherlands, 11-19

Chapter 23

USING OF COLD PLASMA FOR PURIFICATION OF CHEMICALLY POLLUTED WATER IN EXTREME PERIOD

Alexander A. Pivovarov, Anna P. Tischenko
Ukrainian State Chemical-Technology University 8 Gagarin Av., Dnepropetrovsk 49005 Ukraine

Abstract: Investigation of the process of cyanide compounds' destruction under plasma action for a number of model solutions and solutions which can be used for arranging the acts of terrorism was carried out. Dependence of the degree of destruction on composition of solutions and initial concentration of cyanides, as well as on power parameters and duration of treatment was established. It was shown that kinetics of cyanide ions' decomposition in model solutions complies with first-order kinetic equation. Plasma treatment results in cyanide compounds breaking up to non-toxic forms, which can be used for neutralization of these solutions.

1. INTRODUCTION

Currently, the scientists and experts from many countries, including Ukraine, are working on development and implementation of principally new technologies for drinking water and wastewater purification based on various physical and chemical methods [1,2]. These technologies are intended to resolve some important environmental and resource conservation problems, such as purification of industrial sewage from galvanic, electrochemical and other hazardous chemical processes, purification of municipal and household sewage, purification of sewage from organic compounds productions and oil-refineries, purification of microbiological and radioactive sewage, activation of the chemically pure water and many others. Besides, the problem of deep purification of sewage from heavy metals, inorganic and organic admixtures, including diluted solutions, has not been solved yet. Solution of above-mentioned problems will help the citizens to adapt to the modern complex ecological conditions and increase their living standards. During the last ten years, significant progress has been achieved in development of new methods of water treatment based on electrochemical and plasma-chemical principles.

C. Dishovsky et al. (eds.), Medical Treatment of Intoxications and Decontamination of Chemical Agents in the Area of Terrorist Attack, 203–212.

The most perspective are the methods based on application of non-equilibrium low-temperature plasma [3]. They have a number of advantages: smaller dimensions of the equipment, opportunity to automate both the process and quality control of processed environment, low involvement of human resources, opportunity to use new solutions, though poorly investigated, but having useful potential and properties. The basis for the process is the contact plasma discharge on the surface of a liquid phase formed between an electrode in gas phase and surface of the liquid, in which the second electrode is immersed.

Cyanide is a powerful and rapid-acting poison [4]. Hydrogen cyanide has been used in gas-chamber executions and as war gas. Exposure to large amounts of cyanide can be deadly. Cyanide occurs most commonly as hydrogen cyanide in water although it can also occur as the cyanide ion, alkali and alkaline earth metal cyanides (potassium cyanide, sodium cyanide, calcium cyande), relatively stable metallocyanide complexes $[Fe(CN)_6]^{-3}$, moderately stable metallocyanide complexes (complex nicel and copper cyanide), or easily decomposable metallocyanide complexes (zinc cyanide $[Zn(CN)_2]$, cadmium cyanide $[Cd(CN)_2]$. Hydrogen cyanide and cyanide ion combined are commonly termed free cyanide. In process solutions of industrial enterprises, which are using these solutions in galvanic production or hydrometallurgy, this kinetics is distinguished by more complicated character, owing to various speed of decomposition of free cyanide and cyanide-metal complexes. As distinct from the known methods of cyanide neutralization, plasma-chemical treatment allows to break up both high and low concentrations of cyanide ions to the level of maximum allowable concentration and below.

2. EXPERIMENTAL

Experiments were conducted with the use of compact laboratory unit including systems of power supply, vacuumization and thermostating of the reactor of discrete action. This is achieved by positioning two or more pairs of the unlike electrodes into a water layer of between 30 and 100 mm (Fig. 1) on the opposite sides of the "water-air" border at a distance of between 4 to 15 mm from such level accordingly, said electrodes being made of a material which does not have any catalytic effect on cyanide solution, exposing the water to GDP, with voltage at each pair of electrodes being 500 - 1500 V, a current being 25 – 150 mA, the temperature being below the natural boiling point of water and pressure being between $1.5 \cdot 10^4$ - $5 \cdot 10^4$ Pa.

Further investigations proved an assumption stated before, concerning destruction of cyanide complexes and cyanide ions under the action of non-equilibrium plasma [5]. A number of experiments on destructing cyanide ions were conducted both in model solutions and technological solutions containing complex cyanides of various metals (composition of the solutions is given in Table 1.).

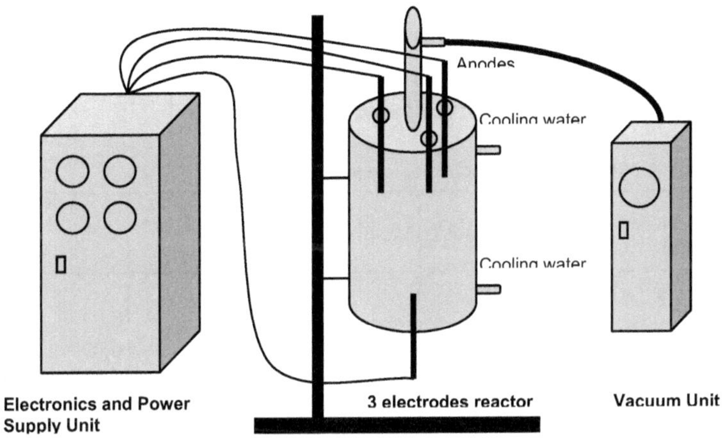

Figure 1. Laboratory setup for activated water production.

Cyanide content was controlled using argentometer and spectrophotometer with barbituric acid and pyridine by means of optimized method [6].

Fig. 2(a) shows dependencies of KCN content in model solutions on time of plasma treatment. They were prepared by means of dilution of the parent solution, containing 21% KCN and 42% KOH. It was shown that at low concentrations of KCN (0,01 - 0,10 %), complete neutralization of solutions was achieved in 2÷8 min. Under cyanide content of 0,3÷1,0 %, this time makes 30 minutes. Solutions with KCN concentration of 1,30 % are not neutralized completely within this time.

Fig. 2(b) represents similar dependencies for technological solutions. Solutions were obtained by means of cyanidation of specified quantities of one metal (Au, Ag, Cu, Zn, curves 1'-6'), or the ore concentrate containing all the above stated metals (curve 7'). Figures prove that process of cyanide destruction is determined by the time of plasma action on the solution. For technological solutions, time of treatment required for complete destruction of cyanide ions depends on composition of the solution. The more complex is the composition, the longer time is required for complete degradation of cyanides. Character of the curves is changed as well.

Calculations show that curves of cyanide ions destruction in the process of plasma treatment for model solutions can be described by first-order kinetic equation $k_l = \frac{2,303}{\tau} \cdot \lg \frac{a}{a-x}$, (1)

where k_l – speed constant;

a – concentration of the substance at the starting moment of time $\tau = 0$;

x – decrease in concentration of the substance on the expiry of time τ.

This conclusion is made as based on the calculation of speed constants for various values of τ for each curve (Table 2).

Table 1. Composition of model and technological solutions.

Solution	C_{KOH}, mol/l	C_{KCN}, mol/l	Metal	C_{Me}, mg/l	k, min^{-1}
1	0,27	0,62	-	-	0,07
2	0,20	0,47	-	-	0,07
3	0,14	0,33	-	-	0,14
4	0,08	0,18	-	-	0,23
5	0,04	0,10	-	-	0,34
6	0,02	0,04	-	-	0,56
7	0,01	0,02	-	-	0,72
1’	0,086	0,148	Au	2860	-
2’	0,057	0,098	Au	710	-
3’	0,032	0,055	Au	1430	-
4’	0,037	0,064	Ag	1420	-
5’	0,022	0,038	Cu	1000	-
6’	0,022	0,047	Zn	640	-
7’	0,039	0,067	Au, Ag, Cu, Zn	*	

*Accordingly, 140, 100, 300 and 600 mg/dm^3

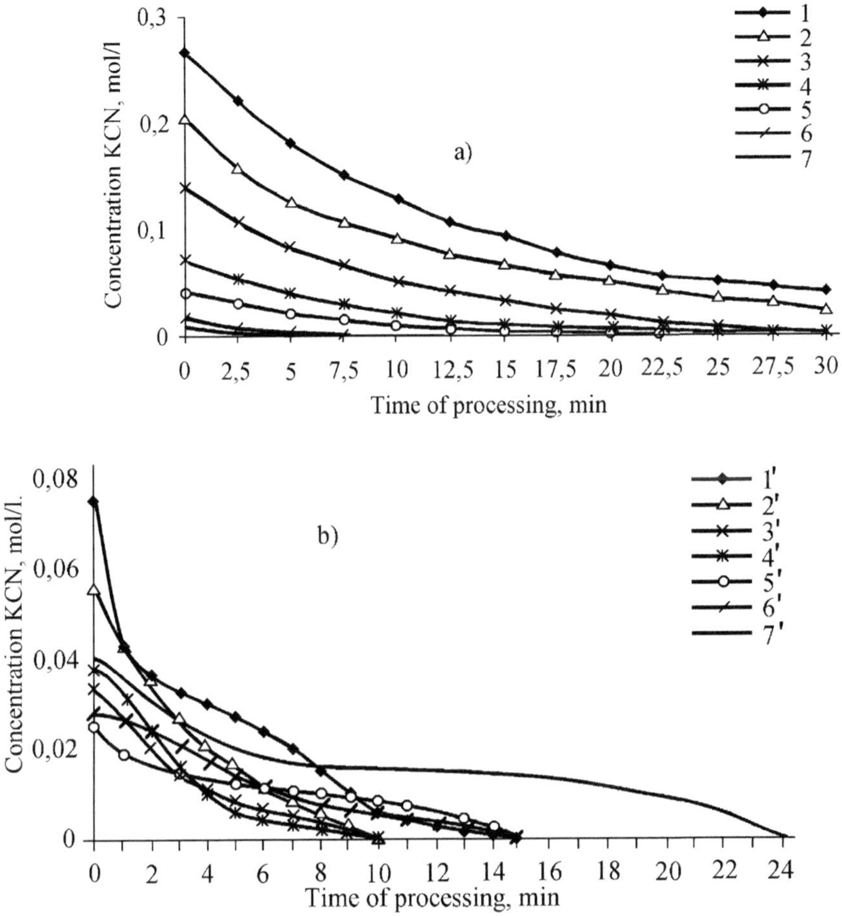

Figure 2. Change in KCN content depending on duration of plasma-chemical treatment of model (a) and technological (b) solutions containing complex compounds: Au (1', 2', 3'); Ag (4'); Cu (5'); Zn (6') и Au, Ag, Cu, Zn (7').

Table 2. Speed constants (continuation).

t, min	$\frac{2,303}{\tau}$, min	C_{KCN}, mol/l	$\frac{a}{a-x}$	$k_l = \frac{2,303}{\tau} \cdot \lg \frac{a}{a-x}$, min^{-1}
1	2	3	4	5
a = 0,2670 mol/l				
5	0,4600	0,1880	1,4262	0,071
10	0,2300	0,1340	1,6772	0,071
15	0,1535	0,1920	2,9000	0,071
20	0,1151	0,1610	4,3500	0,073
25	0,0921	0,1510	5,4374	0,070
30	0,0767	0,1420	6,4444	0,062

a = 0,2030 mol/l				
5	0,4600	0,1410	1,4347	0,072
10	0,2300	0,1000	2,0300	0,071
15	0,1535	0,0710	2,8390	0,070
20	0,1151	0,0520	3,8820	0,068
25	0,0921	0,0350	5,7390	0,069
30	0,0767	0,0210	9,4280	0,074
a = 0,1410 mol/l				
5	0,4600	0,0890	1,5860	0,092
10	0,2300	0,0520	2,7060	0,099
15	0,1535	0,0300	4,6000	0,100
20	0,1151	0,0170	8,3640	0,106
25	0,0921	0,0077	16,4000	0,112
30	0,0767	0	-	-
a = 0,0769 mol/l				
5	0,4600	0,0370	2,0833	0,147
10	0,2300	0,0180	4,1667	0,143
15	0,1535	0,0110	6,6667	0,127
20	0,1151	0,0070	10,000	0,115
25	0,0921	0,0030	25,000	0,129
30	0,0767	0	-	-
a = 0,0422 mol/l				
2	1,1515	0,0270	1,5220	0,210
5	0,4606	0,0100	3,9140	0,210
10	0,2303	0,0030	13,7000	0,260
15	0,1525	0,0020	27,4000	0,220
20	0,1151	0	-	-
a = 0,0170 mol/l				
1	2,303	0,0120	1,3750	0,320
2	1,1515	0,0080	2,0000	0,370
5	0,4606	0,0030	5,5000	0,340
8	0,2979	0	-	-

Linearity of lgC dependence on τ (Fig. 3) proves the first order of the reaction. However, values of the constant of pseudo-first order are varying in time for various initial concentrations of cyanide ions and are actually the same for large enough initial concentrations only (0,2030 and 0,2670 mol/l of KCN).

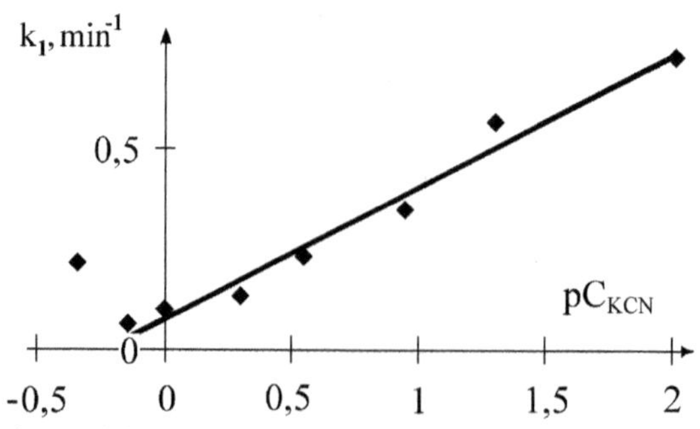

Figure 3. Dependence of the constant of reaction speed on the initial KCN concentration.

It is known that correlation between the substance half-reaction time $\tau_{1/2}$ and its initial concentration C_0 is as follows:

$$\tau_{1/2} \approx C_0^{(1-m_0)}, \qquad (2)$$

where m_0 is kinetic order of the reaction with regard to dissolved substance.

Proportionality of dependence of half-reaction time $\tau_{1/2}$ on the initial concentration of KCN (Fig. 4) can indicate that the limiting stage, determining kinetics of the process, consists in generation of free radicals in the system. Taking into account that with the increase in the initial concentration KCN is present in larger excess as compared with radicals

($H^\bullet$, $OH^\bullet$, HO_2) being generated, zero order of reaction with respect to dissolved substance is to be expected. For treatment of solutions with the initial concentration of KCN, exceeding 0,2030 and 0,2670 mol/l, $\tau_{1/2}$ values are the same. If at low concentrations a stage of radicals' formation of water is significant, direct plasma action on molecules and ions of dissolved substance has greater impact at higher concentrations. Therefore, the order and mechanism of reaction can change.

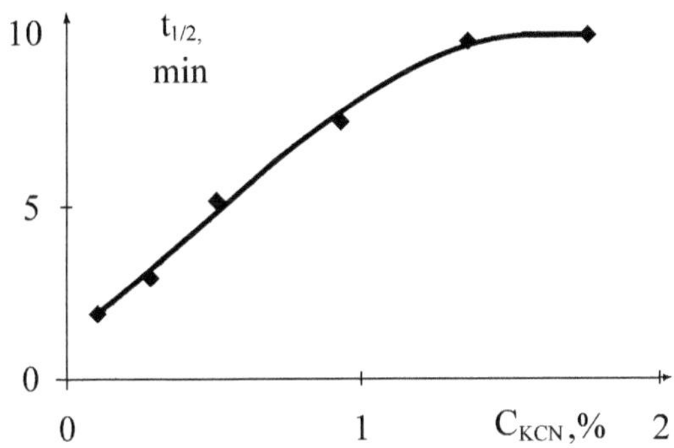

Figure 4. Dependence of half-reaction time $\tau_{1/2}$ on the initial KCN concentration.

In the event of technological solutions, curves of cyanide destruction have horizontal sections which occurrence is caused by various speed of degradation of free cyanide and cyanide, combined in complex with metal.

Determination of the mechanism of chemical transformations in solutions containing cyanide under the action of non-equilibrium plasma represents a complicated task. Plasma action results in generation of radicals, ions, excited molecules of water, secondary electrons and other active particles, which contribute to oxidation-reduction reactions running in the solutions. Due to recombination of HO and HO_2 radicals, highly reactive hydrogen peroxide, also taking part in the reactions, is formed. Obtained dependencies of cyanide ion and hydrogen peroxide concentrations, as functions of time of non-equilibrium plasma treatment (Fig. 5, 6), prove their active interaction. Fig. 5 shows that the larger initial KCN concentration is in the solution, the later H_2O_2 presence in the system is fixed. It may be connected with the fact that H_2O_2 formed in the process of plasma treatment is consumed for interaction with KCN. Curves in the Fig. 6, showing dependence of cyanide destruction efficiency on current strength, allow to draw the similar conclusion.

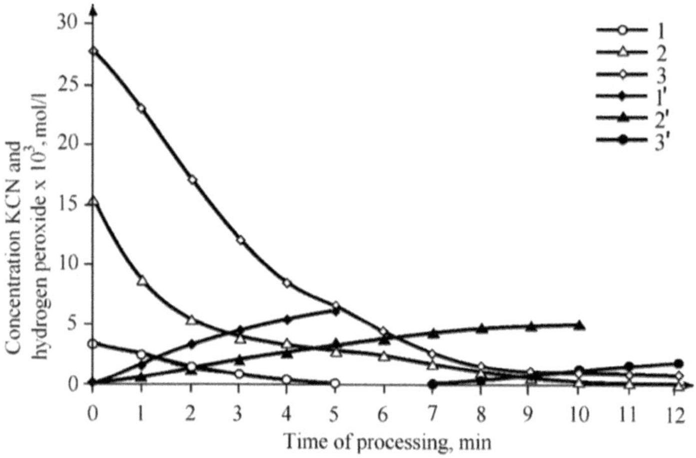

Figure 5. Change in content of KCN (1, 2, 3) and hydrogen peroxide (1', 2', 3') depending on time of plasma treatment in model solutions with various KCN content: 3,3 mol/l - 1, 1'; 15,4 mol/l - 2, 2'; 27,7 mol/l - 3,3'.

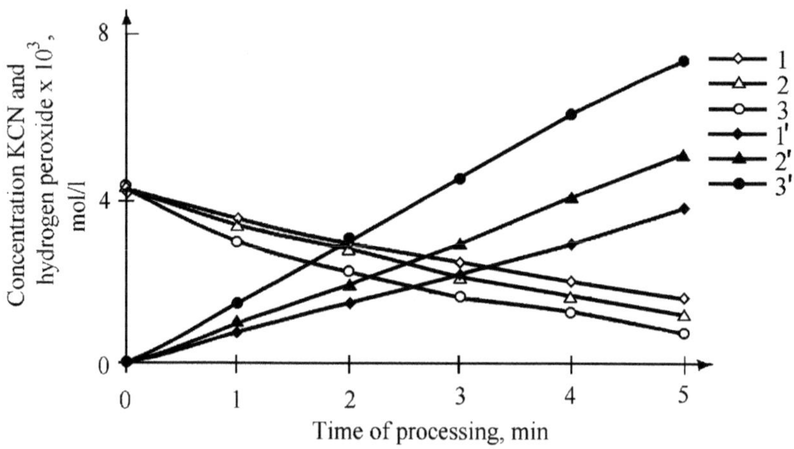

Figure 6. Change in content of KCN (1, 2, 3) and hydrogen peroxide (1', 2', 3') in model solution depending on time of plasma treatment under various current strength: I = 60 мA – 1, 1'; I = 75 мA – 2, 2'; I = 115 мA – 3, 3'.

Formation of HO and HO_2 particles is the direct result of electric discharge; therefore, quantity of the particles formed should be functionally connected with electric parameters of plasma flow. This relation was established with the help of regression analysis using the least-squares method, as based on the experimental measurements of hydrogen peroxide generated in the course of solutions' treatment with plasma. This dependence is expressed by the equation:

$$C = a+b\cdot\tau+c\cdot I+d\cdot\tau\cdot I, \qquad (3)$$

where: C – hydrogen peroxide concentration;

τ - plasma exposure time;

I – set current strength;

a = -4; b = -1,3; c = 0,084; d = 0,424 – calculated constants;

S_{ocm} = 1580 – residual sum of the deviation squares.

When choosing an equation, the following variants were considered:

linear $C= a + b\cdot\tau +c\cdot I$, $S_{ocm} = 5402$,

and quadratic $C = a + b\cdot\tau +c\cdot I + d\cdot\tau^2 +e\cdot I^2$, $S_{ocm} = 5165$.

Equation where the values of current strength and time of plasma exposure are correlated between each other is the most accurate in describing the dependence. This choice is proved by the least value of exact amount - deviation squares $S_{ост}$ for the given equation [7].

The paper [8] includes results of investigating electron mechanisms of the impact of active particles, radicals, hydrated electrons artificially generated by plasma on the behavior of cyanide complexes of zinc in water solutions. The above investigation was conducted using quantum chemistry methods. Quantum-chemical calculation of electron structure of the complexes $Zn(CN)_4^{2-}\cdot 4H^+\cdot 2OH^-$ with complete optimization of all geometric parameters [9] was performed.

Analysis of calculation results has shown profound reconstruction of links between interacting centers that occurred in the chosen model. The most significant changes are observed in the first coordination sphere of central atom. In the process of interaction of $Zn(CN)_4^{2-}$ complex with H^+ and OH^- particles, there is a tendency to destruction of the first coordination sphere of Zn atom with subsequent forming $Zn(OH)_2$ complex and four molecules of HCN. Further HCN degradation in water solution under plasma action is running in accordance with the pattern below:

$$CN^- + 2OH^- \rightarrow CNO^- + H_2O + 2e, \qquad (4)$$

$$2CNO^- + 4OH^- \rightarrow 2CO_2 + N_2 + 2H_2O + 6e, \qquad (5)$$

thus ensuring complete decomposition of toxic cyanide.

3. CONCLUSION

Plasma-chemical sample preparation allows breaking-up of cyanide compounds to non-toxic forms, which may be used for purification of technological solutions and galvanic production wastewaters, posing hazard to people's health in the event they are purposely used by terrorists.

REFERENCES

1. Электрохимическая активация: очистка воды и получение полезных растворов. / В.М.Бахир, Ю.Г. Задорожний, Б.И. Леонов и др. ВНИИИМТ. «Маркетинг Саппорт Сервисиз», 2001, – 176 с.
2. Грановский М.Г., Лавров И.С., Смирнов О.В. Электрообработка жидкостей. - Л.:Химия, 1976. – 216 с.
3. Пивоваров А.А., Сергеева О.В. Физико-химические превращения в водных средах под действием неравновесной плазмы тлеющего разряда //Вопросы химии и химической технологии. 1999. № 2.
C.48-50.
4. Бобков С.С., Смирнов С.К. Синильная кислота. М.: Химия. 1970 – 176 с.
5. Пивоваров А.А., Чмиленко Ф.А., Деркач Т.М., Куксенко А.Н., Майструк А.Н. Плазмохимическое обезвреживание циансодержащих сточных вод // Химия и технология воды. - 1996. -т.18,вып.4. -С. 416-419.
6. Лурье Ю.Ю. Аналитическая химия промышленных сточных вод- -М.: Химия, 1984-
447 с.
7. Львовский Е.Н, Статистические методы построения эмпирических формул. - М.: Высш. шк., 1988. - 238 с.
8. Чмиленко Ф.А., Пивоваров А.А., Каташинский А.С., Деркач Т.М., Куксенко А.Н. Кинетика и механизм разрушения цианидов в промышленных сточных водах неравновесной низкотемпературной плазмой // Плазмотехнология-95: Сб. научн. тр. - Запорожье, 1995. - С. 41-44.
9. Химельблау Д. Прикладное нелинейное программирование. - М.: Мир, 1965, С. 122.

Chapter 24

THE ROLE OF CLINICAL TOXICOLOGY IN MINIMIZING ADVERSE HEALTH EFFECTS RELATED TO CHEMICAL DISASTERS

Givi Katsitadz[1], Manana Juruli[2]
Tbilisi State Medical Academy[1], Geogrian Environmental and Biological Monitoring Association[2] Tbilisi, Georgia

Keywords: precautionary measures, chemical terrorism, chemical enterprises, abandoned military wastes, and disaster preparation

1. INTRODUCTION

Chemical disasters continue to be a major cause of morbidity and mortality. Such disasters include:

- Natural and man-made disasters;
- Terrorist threats or incidents using chemical, biological, nuclear / radiological or large explosive devices;
- Industrial or transportation accidents;
- Use of hazardous chemical waste, including abandoned pesticides and military wastes.

One of the primary consequences of any disaster is its adverse impact on human health and welfare. In order to effectively and efficiently prepare for and respond to possible incidents, public health and medical officials at all levels of government must be prepared to address sudden and unexpected demands for services that may exceed readily available resources. *Conduct and support research and investigations into the cause, treatment, or prevention of a disease or disorders.*

The prevention goals are: safe and secure handling of potential chemical-terrorism agents; safe and secure laboratories for research and testing involving potential terrorism agents; ensuring that food and medical products are secure from potential terrorism agents;

C. Dishovsky et al. (eds.), Medical Treatment of Intoxications and Decontamination of Chemical Agents in the Area of Terrorist Attack, 213–217.

2. CHEMICAL ENTERPRISES

A chemical disaster may occur after damaging of the facilities of the oil, petrochemical and chemical industry. In Georgia, it is enough to point out Zestaponi ferroalloy; steel, nitrate fertilizers, chemical fibers plants and a chemical enterprise "Nitrogen ("Azoti") in Rustavi ("Nitrogen") Available destruction of these chemical plants and manufactures may induce massive releases of hazardous chemicals into environment and in result, cause large-scale immediate and long-term consequences. Toxic substances, released into environment are heavy metals, such as lead, manganese, copper, arsenic, cadmium and others, organic solvents, such as trichlorethylene, benzene, as well as ammonia, cianides, sulphuric and nitrogen oxides and acids. Many of these chemicals have cumulative effects and may continue to damage humans, animals and plants and poison food and water for years and decades.

3. OIL AND GAS PIPLINES

Oil and gas pipelines may also be target for terrorists. In these cases, accidental oil spills and leakages may occur. Due to the highly flammable nature of crude oil and gas, there could arise a serious hazard associated with fire. Crude oil is rich in sulphur compounds. It contains compounds with various amounts of nitrogen and oxygen, plus salt, trace metals and water. Hydrogen sulfide in the crude oil is a lethal gas. Its high concentration should be suspected in the air. Hydrocarbons in crude oil include mixtures of parasitic, olefin, napthenic, and aromatic types. All those hydrocarbons are also potential health stress agents.

4. CHEMICAL WEAPONS

Public health threats and medical emergencies can ensue from terrorist attacks involving chemical weapons. Information on the existence of the chemical weapon on military bases of former Soviet Union in Georgia was never published in the open press, and, of course, we know nothing about it. According to the open press of Russia, it is considered that they are as following: VX gas, sarin, soman, yperite, lewisite, phosgene, and others. It should be noted that CS and chloropicrin (so-called "Tcheriomukha") were applied in April 1989 against peaceful demonstrators in Tbilisi. Accorfding to the above-stated, it is necessary to suppose that, if chemical weapons were kept at Russian military bases in Georgia, it should be similar types of weapons as stored on bases of the Russian territory.

The Chemical Weapons Convention of the 13.01.1993 is disarmament and arms control treaty whose aims are the prohibition of chemical weapons worldwide and the destruction of existing stocks of chemical weapons.

The Chemical Weapons Convention includes lists of the chemicals it controls:

- Highly effective warfare agents (e.g. soman, sarin, tabun) and certain of their direct precursors.
- Key chemicals, which are used for the production of insecticides, herbicides, flame-retardants, lubricants and pharmaceuticals etc. as well as plastics (e.g. amiton, thiodiglycol).
- Intermediates with wide civilian uses, for example, in the production of insecticides, pharmaceuticals, paints and lacquers, coating agents and lubricants etc. (e.g. phosgene, phosphorus oxygen chloride).

An essential aim of future law-making procedures is to harmonise international recommendations and agreements with national and European regulations.

5. HAZARDOUS CHEMICAL WASTE

It is well known that uncontrolled chemical waste, including waste storage and treatment facilities, former industrial and agricultural enterprises and military facilities can contribute to health and environmental problems.

There is growing evidence that hazardous chemicals can affect the central nervous, reproductive, and gastrointestinal system and cause hematological, mutagenic and carcinogenic effects. They can also influence the activity of the immune system by either augmenting or suppressing its function.

At present most of the pesticide and fertilizer storehouses are in a very bad condition. None of them answer to environmental and sanitary demands. Transportation, usage, distribution and storage of chemicals are not regulated. Hazardous chemicals are in sale everywhere - in the market, on the roads, in the boxes with food products, often without labeling. It is impossible to control chemicals. There is not a special waste disposal.

In accordance with data of 2001, military chemical waste in Georgia consists of rocket fuel (about 1500 tonnes), napalm (912 tonnes), and large amount of chloropicrin. In spite of the fact that they are partially utilized, the rest poses still concern and requires great attention.

As to the process of recycling, it is not so simple. Significant resources (human as well as financial resources) are needed for this purpose. The donor-countries support Russia in this matter.

In Georgia, due to lack of financial resources it is difficult to take appropriate measures for utilization, segregation, recovering and recycling of waste. In most cases, there is a complete absence of routine monitoring for a variety of reasons. These are: old analytical laboratory equipment, absence of reagents, and standards for environmental and biological monitoring.

6. CONCLUSION

Chemical disasters are not very easy to predict and prevent. Depending on the extent of the accident, they may produce the following health manifestations: asphyxiation, central nervous system depression, defeating dermatitis, aspiration pneumonitis, myocardial sensitization and irritability and hepatorenal toxicity.

It might be said that huge social, political, environmental and health problems still confront Georgia. There are not sources to provide decontamination of toxic chemicals left without any control at the old military bases and industrial and agricultural facilities.

The situation at the laboratories also reflects the political and economic situation. Due to lack of funding all of those laboratories are destroyed. Otherwise, the absence of reagents, glassware and the presence of old and unserviceable equipment meant that no monitoring had taken place since 1990. Often there is no electricity (especially in wintertime) and water supply is cut off at intervals. Air conditioning is not available in laboratories. The ventilation system in most laboratories does not work and there is need for repair of the ventilation system. There is need to improve the general laboratory facilities of the laboratories.

One of the primary consequences of any disaster is its adverse impact on human health and welfare. Therefore, public health and medical officials at all levels of government must be prepared to address sudden and unexpected demands for services that may exceed readily available resources.

In order to effectively and efficiently prepare for and respond to possible incidents and to keep consequences of chemical terrorist attacks to a minimum the following actions are recommanded:

- To make an inventory of sources of hazardous chemical waste;
- To provide risk assessment of toxic waste sites;
- To clean up (and redemption) of hazardous waste sites across Georgia;
- To improve environmental and biological monitoring;
- To improve the national legal instruments related to chemical management and chemical weapons;
- To improve the general laboratory facilities with major analytical and computing equipment; provide training to professionals;
- To develop teaching materials and core curricula for public health officials, medical professionals, emergency physicians and emergency department staff, laboratory personnel and other personnel working in

health care facilities, for recognition and identification of potential bioweapons and other agents that may create a public health emergency, and care of victims of such an emergency, recognizing the special needs of children and other vulnerable populations;

Chapter 25

PREVENTION OF GAS SEEPING INTO BUILDINGS THROUGH CONSTRUCTIVE MATERIALS

Rashid A. Khaydarov, Renat R. Khaydarov and Olga U. Gapurova
Institute of Nuclear Physics, 702132, Ulugbek, Tashkent, Uzbekistan,

Abstract: The investigation described in this article examined the method of chemical treatment of walls, floor, ceiling, roof, etc. to prevent gas seeping into buildings. A mixture of organic compounds and catalyst are used to fill micropores inside the concrete or another constructive materials against gases, as well as water molecules. The method allows reducing the coefficient of gas (air, Ar, ^{222}Rn, ^{85}Kr, $^{133,135}Xe$, etc.) permeability in 200 – 400 times. Consumption of the chemicals is 0.2 L/m^2 for gypsum and 0.3-0.4 L/m^2 for concrete and cement and depends on porosity of the materials.

Keywords: chemical terrorism, constructive materials, gas seeping, micropores, penetration, radioactive gases

1. INTRODUCTION

One of possible methods of realization of the terrorist acts is using gases and liquids, which easily permeate through the constructive materials of walls, floor, ceiling, roof, etc. into buildings by the capillary action of the pores. Toxic volatile organic compounds, organic and inorganic gases, radioactive elements, particularly those emitting alpha particles can be used as the dangerous substances. Increased ventilation may help in removing the gases, but can actually increase the gases level by increasing the suction through the pores of concrete. If the gases and liquids are soluble in water and are easily volatilized from it, they can also get by groundwater up to underground structures and penetrate inside through opening and pores in concrete or pushed by hydrostatic pressure.

The purpose of this work is creating a method to reduce concentration of toxic and radioactive gases in homes, buildings, underground structures, tunnels, hangars, garages, bombshelters, etc. This task can be solved by using special chemicals, which fill microcrevices and minute pores inside the concrete or other building materials against gases, as well as water molecules. The lay with filled pores stops the

C. Dishovsky et al. (eds.), Medical Treatment of Intoxications and Decontamination of Chemical Agents in the Area of Terrorist Attack, 219–224.

diffusion of molecules of gases and income through the slab, as well as the walls, floors and ceilings. And in addition, it blocks the other gases pathway - water migration.

There are many chemicals preventing water diffusion in concrete and other constructive materials, but they do not change gas permeability of concrete and gases migrates through them easily. The chemicals intended for prevention of gas seeping through constructive materials must meet following requirements: low price and consumption, high efficiency with reducing the gas permeability 400-500 times, wide range of operation temperature from 1-2 ^{0}C to 40-50 ^{0}C and air humidity from 2-5% to 95-98%, low water consumption.

2. MATERIALS AND METHODS

The following organic compounds have been tested: polyethylhydrosiloxane (PEHS) and its aqueous emulsion, mixture of PEHS and benzene in the ratio 1:1, 25 % solution of alkyltriethoxysilane (ATES) in isopropanol (25% ATES), mixture of PEHS and 25% solution of ATES in isopropanol. Sn-, Ti-, and Cu- organic compounds were tested as catalysts.

The tested samples of concrete, cement and gypsum had height and diameter of 15-20 mm and 30 mm, respectively. The samples in a vertical position were treated by chemicals by means of a spray.

The device for examination of gas permeability of samples of constructive materials has been constructed. The device consists of compressor 1, receiver 2 with manometer 3, gates 4,5, holder 6 of concrete samples 7, flow meter of air 8 and vessel with gas 9 (Fig.1). The gate 5 adjusted the value of the gas pressure. In all experiments the pressure of gases in the receiver was 110 kPa. Air, Ar and ^{222}Rn with concentration of 100 pCu/L in air were used as the gases.

The Coefficient of gas permeability K was determined by the following formula

$$K = \Delta V\, d / S\, \Delta t\, \Delta p, \quad (1)$$

where Δt is the time of passing the gas through the sample, ΔV is the volume of gas passed through the sample, Δp is a pressure drop of gas in the sample, S is an area of the sample, d is a thickness of the sample. In these experiments we measured the relation R between the coefficients of gas permeability for chemically non- treated samples K_0 and treated samples K:

$$R = K_0/K. \quad (2)$$

The gas permeability of the constructive materials after treatment by chemicals was examined. Influence of types of concrete, cement and gypsum, preliminary treatment by different chemicals, time between treatments by organic compounds, humidity of constructive materials, time between preparation of chemicals and treatment of materials (aging of chemicals), time between treatment of materials and testing (aging of treated materials) were investigated.

The concentration of ^{222}Rn in air was determined with a radon measurement detector. The detector allows realizing continuous radon monitoring. It consists of an electronic unit and a scintillation cell. The electronic unit contains power supply, amplifier, discriminator, timer, counter, and indicator. The scintillation cell contains the zinc sulfide scintillator, photomultiplier, preamplifier, high voltage power supply and chamber with a volume of 200 mL over the scintillator. This chamber is filled with the gas to be analyzed. The air is either pumped or diffuses into the scintillation cell. The scintillation count is processed by electronics, and radon concentrations for predetermined intervals are stored in the memory of the device.

3. RESULTS AND DISCUSSION

In the system ATES - isopropanol a re-etherification of the following type takes place:

$$AlkSi(OEt)_3 + (3\text{-}n)ROH \rightarrow AlkSi(OEt)_n(OR)_{3-n} + (3\text{-}n)EtOH$$

where n = 0-2, R is C_3H_7 . $AlkSi(OR)_3$, $AlkSi(OEt)(OR)_2$, $AlkSi(OEt)_2(OR)$ are the products of the re-etherification.

Products of this reaction are easily hydrolyzed even by air moisture:

$$AlkSi(OR)_3 + 3\,H_2O \rightarrow AlkSi(OH)_3 + 3ROH$$

with formation of the intermediate compounds $AlkSi(OR)_2OH$ and $AlkSi(OR)(OH)_2$. When PEHS is introduced in the hydrolyzed solution, polymeric networks of linear, cyclical and ramified structures with different properties are forms, for example:

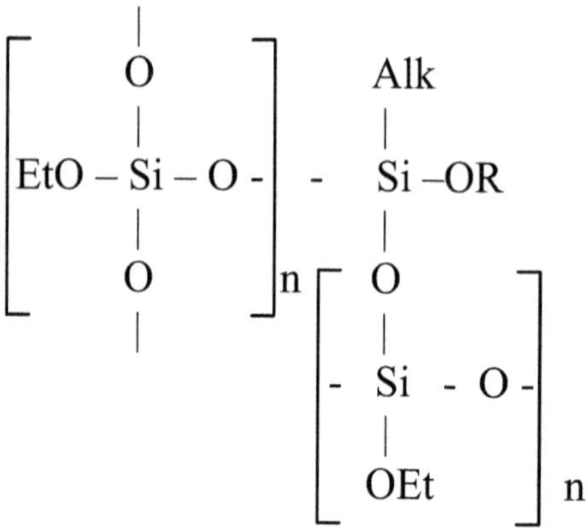

Hydrolysis of PEHS in early stages also takes place. Then the products of hydrolysis interact with ATES. The preferable formation of polymeric network type depends on methods of the mixture preparation, treatment of the materials and etc. Besides the reaction with $Ca(OH)_2$ takes place and as the result the polymer is fixed strongly in the treated materials.

Experiments have shown the following results. The depth of penetration of the organic compound without catalysts is about 10-14 cm. However, the polymerization time is very long (about 2-3 days) and consumption of the compound is substantial (about 1-2 L/m^2). The use of catalysts reduces the duration of the polymerization process to 1-3 hours. In this case the depth of penetration of chemicals into constructive materials is about 5-6 mm and consequently the consumption of organic compounds decreases. The treated materials have hydrophobic properties. The relation R does not depend on humidity of the surface of materials.

Dependence of the relation R with the number of layers is given in Table 1. The best type of catalyst is the Sn-organic compound. The relation R for concrete increases slowly after the first layers and very rapidly after the 4th and 5th layers. This is explained by the high porosity of concrete. In the case of cement and gypsum R increases rapidly after the 2nd layer. The total consumption of chemicals is about 0.4 L/m^2 for concrete (but it depends on type of concrete), 0.3-0.4 L/m^2 and 0.2-0.3 L/m^2 for cement and gypsum, respectively. There is no a great difference between air, Ar and radon permeabilities.

Dependence of the relation R with the time after chemical treatment of construction materials is given in Table 2. After the 4th day the relation R increases very slowly, about 5% within 7 months.

4. CONCLUSION

The test results of these investigations demonstrate the efficacy of the described method of chemical treatment of walls, floor, ceiling, roof, etc. to prevent gas from seeping into buildings. The method allows reducing the coefficient of gas (air, Ar, ^{222}Rn, ^{85}Kr, $^{133,135}Xe$) permeability 200 – 400 times. Consumption of the chemicals is 0.2 L/m^2 for gypsum and 0.3-0.4 L/m^2 for concrete and cement and depends on porosity of the materials. This method can be used for prevention of seeping dangerous gases through the constructive materials.

Table 1. Dependence of R on the number of layers.

Gas	Building material	Type of catalyst	The number of layers				Total consumption of chemicals, L/m^2
			2	3	4	6	
Air	Concrete	Sn-organic compound	2.1	3.5	110	420	0.402
		Ti-organic compound	2.0	3.1	80	320	0.560
		Cu-organic compound	1.2	2.4	45	100	0.610
	Cement	Sn-organic compound	2.5	47	120	>500	0.360
		Ti-organic compound	2.3	34	98	390	0.410
		Cu-organic compound	1.8	12	59	160	0.520
	Gypsum	Sn-organic compound	2.8	90	410	>500	0.270
		Ti-organic compound	2.6	49	380	440	0.330
		Cu-organic compound	2.2	35	280	360	0.410
Ar	Concrete	Sn-organic compound	2.0	3.4	105	400	0.400
	Cement	Sn-organic compound	2.3	45	115	>500	0.350
	Gypsum	Sn-organic compound	2.7	86	400	>500	0.260
^{222}Rn	Concrete	Sn-organic compound	2.1	3.5	110	430	0.400
in air	Cement	Sn-organic compound	2.4	46	120	>500	0.340
	Gypsum	Sn-organic compound	2.7	88	420	>500	0.260

Table 2. Dependence of R on the time after treatment.

Gas	Building material	Time after treatment, days					
		4	18	28	62	154	210
Ar	Concrete	420	420	430	430	440	440
	Cement	>500	>500	>500	>500	>500	>500
	Gypsum	>500	>500	>500	>500	>500	>500
^{222}Rn	Concrete	430	440	440	450	450	450
	Cement	>500	>500	>500	>500	>500	>500
	Gypsum	>500	>500	>500	>500	>500	>500

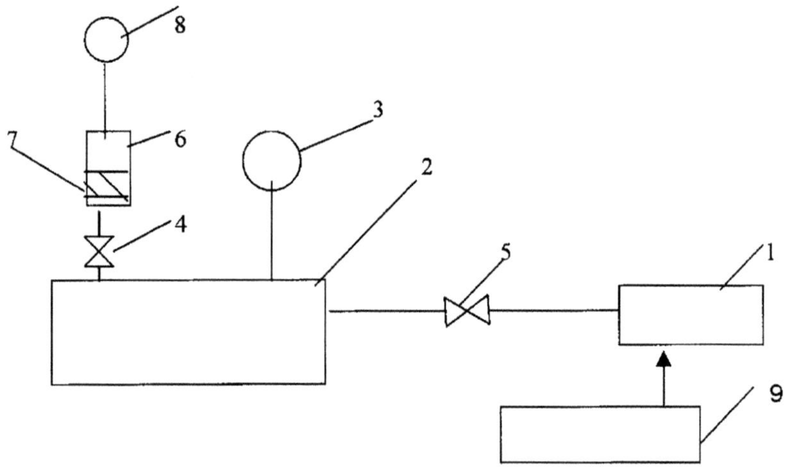

Figure 1. The scheme of the device for examination of gas permeability of building samples: compressor 1, receiver 2 with manometer 3, gates 4 and 5, holder 6 of building samples 7, flow meter of air 8, vessel filled with gas 9.

Chapter 26

BIOLOGICAL METHOD FOR THE WATER, FOOD, FODDERS AND ENVIRONMENT TOXIC CHEMICAL MATERIALS CONTAMINATION INDICATION

L.I. Pozdnyakova, V.P. Lozitsky, A.S. Fedchuk, I.N. Grigorasheva, Y.A. Boshchenko, T.L. Gridina, S.V. Pozdnyakov
Ukrainian I.I.Mechnikov Anti Plague Research Institute. Odessa, Ukraine

Abstract: Early indication and identification of unknown chemical and biological agents play the leading role in the system of countermeasures on liquidation and minimization of the connatural, technogenic and deliberate extreme situations. . These necessitate the use of the special laboratory chemical/toxicological, microbiologic, immunological methods of investigation using special hi-tech equipment and specialized laboratories. It is necessary from several hours up to several days for an integrated analysis. Development and introduction of new and simple methods of rapid indication are actual in present. We have proposed method of nonspecific indication of toxic pollutions possible at the various situations including leakages of toxic materials during production of the ordinary chemical compoiuiua and their hit into the objects of the environment, water and food, and also owing to the disasters and deliberate use of chemicals as agents of terrorism. As the ground of this method is the Colpoda steinii (Infuzoria Family) dry culture application, actuary used for the ecologic/toxicological external environment investigations. This method has been applied earlier for an food toxicity evaluation in animal husbandry industries. We have developed the "know-how" for the diagnostic test-system (culture Colpoda steinii) and started production in the Ukraine (GOST 13496.7-97). Method is based on studying of various parentage toxic materials, including chemical compounds, influence on alive biological objects at the level of the integrated organism. Toxicity of the contaminated objects determine by their influence on motility and vital activity of the Colpoda steinii culture. Time from the beginning of the tested probes (water, various surfaces washouts, suspension and solutions of a food products, etc.) influence is the criterion of the toxicity definition. Probes at which addition drop of various degree motility or even colaps of all culture is observed, count as contaminated. Probes with use of which the Colpoda steinii culture destruction comes within 10 minutes of observation consider **highly toxic,** during till 3 hours - **slightly toxic** and if in 3 hours of all culture remains mobile - nontoxic. Method is simple in use, allows to receive confirmation of the pollution and estimate preliminary contamination degree in a short time (from 10 minutes till 3 hours) in

C. Dishovsky et al. (eds.), Medical Treatment of Intoxications and Decontamination of Chemical Agents in the Area of Terrorist Attack, 225–230.

the inveronment without use of special expensive equipment. The prelimenary estimation can be made before reception of the conclusion about character of the toxic agent and its identification, and the first clinical signs of a lesion. Besides, this method allows to reveal presence residual polutions and to estimate efficacyy of the carried out decontamination.

Keywords: biological model; control; decontamination; indication; toxic chemical material

In the system of the counter measures on liquidation and minimization of the connatural, technogenic and deliberate extreme situations consequences early indication of toxic agents play the leading role. These needs use of the special laboratory chemical/toxicological, microbiologic, immunological methods of investigation with use of special hi-tech equipment and specialized laboratories. It is necessary from several hours up to several days for an integrated analysis. Development and introduction of new and simple methods of rapid indication are actual in present. Now dilating a wide range of rapid indication methods of toxic substances and also search and selection of adequate experimental models are necessary.

The problem of toxic subjects detection in the tested objects can be solved by two options: chemical analysis, for revealing separate toxics, or their products, and biotesting with the result of the tested samples toxicity degree indication without identification of the agent. Qualitative and quantitative chrmical/analytical methods allow with the higher accuracy and, in some cases, rapidly detect presence of the separate toxics or their products in the tested objects. It is important for the regular detection of the different pollutions of any agents in the tested objects.

However, these methods do not allow to judge the full opportunity of alive creatures reaction after influence of the tested toxics agent, which presence was not supposed during examination.

Besides, it is difficult to predict result of the combined action of two and more toxic products.

For the evaluation of the connatural waters toxicity, industrial waste, forages and other objects of the environment various biological objects are in use. Biotesting enables to determine the toxicity of the tested objects with the high degree of the reliability. By the present time a line of the biotesting methods are developed and applied.

As biomodel are used representatives of various systematic groups: mammalian, fishes, ракообразные, supreme plants, algas, the elementary creatures, funguses, bacteria, cellular cultures. Majority of these methods application demand for testing long time and frequently expensive equipment.

One of the most simple in application and an accessible test-object for biotesting are Infusorians. Incentive motives for the infusorians use in ecotoxic testing are the following [1, 2]:

1. - cosmopolitan distribution facilitates comparisons of test results in geographically different regions; 2. - problems of scale are diminished; 3. - replicability is as good as or better than for tests with larger organisms; 4. - environmental realism is higher than is the case for tests using larger organisms 5. - using infusorians increases dramatically the number of tests with limited array of larger organisms; 6.- testing with infusorians is less likely to antagonize animal rights activists; 7. - validation of laboratory tests in field enclosure is facilitates and much less costly.

Advantages of infusorians use in the biotesting are [3]: 1. - infusorians are highly sensitive to the pollutants entering aquatic ecosystems with waste water. They are mostly sensitive to heavy metals (mercury, copper, cadmium, nickel, chromium, zinc, lead). Also, they are sensitive to the toxicants of organic nature (oil, DDT); 2. - using the infusorians we can examine both cell and organism reactions upon the toxicity of effect; 3. - the opportunity of cultivation in a wide range of temperatures allows to use model for experimental research at any time year; 4. - a short life cycle, high speed of reproduction allows to trace the reason upon intoxication during a rather short period in a large number of generations; 5. - using cloning methods it is possible to receive a large number of genetically homogeneous material.

Reports about using of *Colpodidae* in bioassays have been published repeatedly. Infusorians *C.steinii* has all merits like other infusorians as test-objects. However a number of physiological peculiarities distinguish it from row of other infusorians [4]. The conjugation process of this species has not been observed and described in literature for the whole period of investigation since XIX century. *C.steinii* reproduces by palintomic. Genetic constancy of infusorians of this species may be changed only by mutation and clones aging does not occur. *C.steinii* can form resting cysts in unfavorable conditions. Reorganization of macronucleus like reorganization during division happens during encystment. Besides it, part of macronuclear DNA extrusion takes place. Excystment occurs when favorable conditions come. This process goes on 2 h if cysts are moist and more then 8 h if they are dry. In dry state resting cysts can maintain vitality for several years. The last quality allows preparing the infusorians culture for experiments beforehand and keeping it in dry resting cysts state. Excystment is carried out by adding special medium, if it is necessary. Synchronous infusoria culture forms after induced excystment. All cells in the culture are I one stage of life cycle. Genetic system of all *C.steinii* cells in postcyst stage is in the same state as cells after division. We have proposed method of nonspecific indication of toxic pollutions possible at the various situations including leakages of toxic materials during production of the ordinary chemical compounds and their hit into the objects of the environment, water and food, and also owing to the disasters and deliberate use of chemicals as agents of terrorism. As the ground of this method is the *Colpoda steinii* dry culture

application, actually used for the ecologic/toxicological external environment investigations. This method has been applied earlier for food toxicity evaluation in animal husbandry
industries [5].

The technology of dry culture *Colpoda steinii* production is developed adjusted in the Ukraine (the Veterinary Products Company « Renaissanc M »).

Toxicity of the contaminated objects determine by their influence on motility and vital activity of the *Colpoda steinii* culture. Time from the beginning of the tested probes (water, various surfaces washouts, suspension and solutions of a food products, etc.) influence is the ccriterion of the toxicity definition. Probes at which addition drop of various degree motility or even collapse of all culture is observed, count as contaminated. Probes with use of which the *Colpoda steinii* culture destruction comes within 10 minutes of observation consider **highly toxic**, during till 3 hours – **slightly toxic** and if in 3 hours of all culture remains mobile – **nontoxic.** Method is simple in use, allows to receive confirmation of the pollution and estimate preliminary contamination degree in a short time (from 10 minutes till 3 hours) in the environment without use of special expensive equipment.

We studed a toxicity of some medicinal preparations on culture of the primerely trepsinised chicken fibroblasts (TCF) and biotesting with use of *Colpoda steinii* infusorians Results of researches are submitted in the table 1.

Table 1. Results of a toxicity definition of some medicinal preparations on *Colpoda steinii* and TCF.

#	Preparation	The maximal concentration mcg / ml	
		Colpoda steinii	TCF
1	Decametaximun	3	4
2	Ethonium	2,5	35
3	Acyclovir (Zovirax)	>125	>125
4	Amben	>200	>200
5	E-aminocapronic acid	>40000	8000
6	Sodium Aceminatum	>20000	4500

Relative analysis has shown, that the results received by both methods, completely correlate:

Ethonium and Decametaximun have shown rather high toxicity at both methods of testing whereas Zovirax (Acyclovir), E-aminocapronic acid, Ambenum and Sodium Aceminatum had a hypotoxicity at their influence on cells monolayer, and *Colpoda steinii* infusorians culture.

Thus, from the moment of the beginning for biotesting with *Colpoda steinii* preparation and results reception passed 16 – 19 hours, this work did not need a sterile conditions and was carried out by one employee.

Definition of a toxicity with culture of the primerely trepsinised chicken fibroblasts (TCF) included the following stages: - incubation of chicken embryo within 11 day; - tripsinisation and cultivation of a TCF cells monolayer – 2 days; - contact of the tested preparation with the TCF cells monolayer an and the count of a toxicity by the citopathogenic action (CPA)– 1-5 day. Hence, definition of a toxicity on cell culture borrowed 14-18 days and at the last two stages demanded strict sterility.

Thus, it is possible to make the conclusion, that the offered by us method can be used for the toxicity definition in a various objects.
This method is objective, is more simple at execution and demands considerably shoter time and labour expenses.

At the environmental objects toxicity definition with offered method will allow during short time to estimate the toxicity degree of of various objects after eco-disasters or acts of terrorism, with use of the toxic agents. Besides, suggested method does not demand a sterility and allows to use materials from the objects of the external environment, which are not sterile.

The preliminary estimation can be made before reception of the conclusion about character of the toxic agent, its identification, and the first clinical signs of a lesion. Besides, this method allows to reveal presence residual pollutions and to estimate efficacy of the carried out decontamination.

Poposed by our groupe method can be used not only for fast детексии toxic agents, but also for the research purposes in a complex with other biological both physiological methods and models. Among them is «Method for evaluation of the electrocenethic motility of the buccal epithelium cells» intended for the evaluation of the physiological status of organism and an membrane-tropic, toxic or allergic action of medicinations and toxins. In addition – usefull for the screenings "in the field"[6].

For carrying out fundamental research in this direction, the model (biotestsystem), based on the cocultivation of the peripheric blood cells of the Humans with the various infection diseases agents such as Diphtheria, Tularemia, Tuberculosis, Cholera, HIV-infection is developed [7].

With the use of this model is possible to estimate a toxicity degree, the virulence lavel of the agents, and also to estimate bacteriostatic and possible toxic medications manifistations, at their presence in the experimental system.

REFERENCES

1. Cairns, J.,Jr. Myths impeding the utilization of infusoria in ecotoxicological toxicity testing / Infusoria in bioassays, Saint Petersburg, 1998.PP.11-21.
2. Cairns, J.,Jr. Environmental monitoring for sustainable use of the planet / Population and Environment, 1997, v. 18, # 5,PP. 463-471.
3. Bakaeva E. N. The reasons for usage of unicellular in biotesting / Infusoria in bioassays, , Saint Petersburg, 1998.PP.26-27

4. Vinokhodov D.O., Vinokhodov V.O. *Colpoda steinii* as a test organism / Infusoria in bioassays, Saint Petersburg, 1998.PP.85-87.
5. Vinokhodov D.O. Toxicological studies fodders with infusoria using / Saint Petersburg, 1995, 80 p.
6. G.F.Biloklitska, M.A.Novikova, M.A.Volik, L.I.Pozdnyakova. Test-system for the organism physiologic status evaluation in stomatology by EKB and KBE / Proceedings of the VI Congress of the World Federation of the Ukrainian Doctors Societies. Odessa, 1996, P.193-195.
7. Y.A. Boshchenko, A.G.Stopchamskaya, L.I.Pozdyakova. Modelling "in vitro". Interactions of the high pathogenic agent – Fr.tularensis with Human immune competent cells / 4th SISPAT Singapore International Symposium on protection against toxic substances / Singapore, 2004, P.145.

INDEX

A

B

C

D

E

F

G

H

I

L

M

N

O

P

GPSR Compliance

The European Union's (EU) General Product Safety Regulation (GPSR) is a set of rules that requires consumer products to be safe and our obligations to ensure this.

If you have any concerns about our products, you can contact us on ProductSafety@springernature.com

In case Publisher is established outside the EU, the EU authorized representative is:

Springer Nature Customer Service Center GmbH
Europaplatz 3
69115 Heidelberg, Germany

Batch number: 10583998

Printed by Printforce, the Netherlands